Climate Change

ISSUES

Volume 151

Editors

Cobi Smith and Lisa Firth

First published by Independence
The Studio, High Green
Great Shelford
Cambridge CB22 5EG
England

© Independence 2008

British Library Cataloguing in Publication Data
Climate Change – (Issues Series)
I. Smith, Cobi II. Series
363.7'3874

ISBN 978 1 86168 424 0

Printed in Great Britain
MWL Print Group Ltd

Cover
The illustration on the front cover is by
Angelo Madrid.

CONTENTS

Useful information for readers

Dear Reader,

Issues: Climate Change

The controversy surrounding climate change continues to rage – are man's actions rather than a natural phenomenon responsible for climate change, and if we alter our behaviour will it make any difference? This book looks at the debate about global warming, as well as at the causes and effects of climate change and their implications for the future.

The purpose of *Issues*

Climate Change is the one hundred and fifty-first volume in the **Issues** series. The aim of this series is to offer up-to-date information about important issues in our world. Whether you are a regular reader or new to the series, we do hope you find this book a useful overview of the many and complex issues involved in the topic.

Titles in the **Issues** series are resource books designed to be of especial use to those undertaking project work or requiring an overview of facts, opinions and information on a particular subject, particularly as a prelude to undertaking their own research.

The information in this book is not from a single author, publication or organisation; the value of this unique series lies in the fact that it presents information from a wide variety of sources, including:
⇨ Government reports and statistics
⇨ Newspaper articles and features
⇨ Information from think-tanks and policy institutes
⇨ Magazine features and surveys
⇨ Website material
⇨ Literature from lobby groups and charitable organisations. *

Critical evaluation

Because the information reprinted here is from a number of different sources, readers should bear in mind the origin of the text and whether the source is likely to have a particular bias or agenda when presenting information (just as they would if undertaking their own research). It is hoped that, as you read about the many aspects of the issues explored in this book, you will critically evaluate the information presented. It is important that you decide whether you are being presented with facts or opinions. Does the writer give a biased or an unbiased report? If an opinion is being expressed, do you agree with the writer?

Climate Change offers a useful starting point for those who need convenient access to information about the many issues involved. However, it is only a starting point. Following each article is a URL to the relevant organisation's website, which you may wish to visit for further information.

Kind regards,

Lisa Firth
Editor, **Issues** series

** Please note that Independence Publishers has no political affiliations or opinions on the topics covered in the **Issues** series, and any views quoted in this book are not necessarily those of the publisher or its staff.*

ISSUES TODAY
A RESOURCE FOR KEY STAGE 3

Younger readers can also now benefit from the thorough editorial process which characterises the **Issues** series with the launch of a new range of titles for 11- to 14-year-old students, **Issues Today**. In addition to containing information from a wide range of sources, rewritten with this age group in mind, **Issues Today** titles also feature comprehensive glossaries, an accessible and attractive layout and handy tasks and assignments which can be used in class, for homework or as a revision aid. In addition, these titles are fully photocopiable. For more information, please visit the **Issues Today** section of our website (www.independence.co.uk).

Instant expert: climate change

Information from the *New Scientist*

By Fred Pearce

Climate change is with us. A decade ago, it was conjecture. Now the future is unfolding before our eyes. Canada's Inuit see it in disappearing Arctic ice and permafrost. The shantytown dwellers of Latin America and Southern Asia see it in lethal storms and floods. Europeans see it in disappearing glaciers, forest fires and fatal heatwaves.

Scientists see it in tree rings, ancient coral and bubbles trapped in ice cores. These reveal that the world has not been as warm as it is now for a millennium or more. The three warmest years on record have all occurred since 1998; 19 of the warmest 20 since 1980. And Earth has probably never warmed as fast as in the past 30 years – a period when natural influences on global temperatures, such as solar cycles and volcanoes should have cooled us down. Studies of the thermal inertia of the oceans suggest that there is more warming in the pipeline.

Climatologists reporting for the UN Intergovernmental Panel on Climate Change (IPCC) say we are seeing global warming caused by human activities and there are growing fears of feedbacks that will accelerate this warming.

Global greenhouse

People are causing the change by burning nature's vast stores of coal, oil and natural gas. This releases billions of tonnes of carbon dioxide (CO_2) every year, although the changes may actually have started with the dawn of agriculture, say some scientists.

The physics of the 'greenhouse effect' has been a matter of scientific fact for a century. CO_2 is a green-house gas that traps the Sun's radiation within the troposphere, the lower atmosphere. It has accumulated along with other man-made greenhouse gases, such as methane and chlorofluorocarbons (CFCs).

If current trends continue, we will raise atmospheric CO_2 concentrations to double pre-industrial levels during this century. That will probably be enough to raise global temperatures by around 2°C to 5°C. Some warming is certain, but the degree will be determined by feedbacks involving melting ice, the oceans, water vapour, clouds and changes to vegetation.

Warming is bringing other unpredictable changes. Melting glaciers and precipitation are causing some rivers to overflow, while evaporation is emptying others. Diseases are spreading. Some crops grow faster while others see yields slashed by disease and drought. Strong hurricanes are becoming more frequent and destructive. Arctic sea ice is melting faster every year, and there are growing fears of a shutdown of the ocean currents that keep Europe warm for its latitude. Clashes over dwindling water resources may cause conflicts in many regions.

As natural ecosystems – such as coral reefs – are disrupted, biodiversity is reduced. Most species cannot migrate fast enough to keep up, though others are already evolving in response to warming.

Thermal expansion of the oceans, combined with melting ice on land, is also raising sea levels. In this century, human activity could trigger an irreversible melting of the Greenland ice sheet and Antarctic glaciers. This would condemn the world to a rise in sea level of six metres – enough to flood land occupied by billions of people.

The global warming would be more pronounced if it were not for sulphur particles and other pollutants that shade us, and because forests and oceans absorb around half of the CO_2 we produce. But the accumulation rate of atmospheric CO_2 has increased since 2001, suggesting that nature's

ability to absorb the gas could now be stretched to the limit. Recent research suggests that natural CO_2 'sinks', like peat bogs and forests, are actually starting to release CO_2.

Deeper cuts

At the Earth Summit in 1992, the world agreed to prevent 'dangerous' climate change. The first step was the 1997 Kyoto Protocol, which finally came into force during 2005. It will bring modest emission reductions from industrialised countries. But many observers say deeper cuts are needed and developing nations, which have large and growing populations, will one day have to join in.

Some, including the US Bush administration, say the scientific uncertainty over the pace of climate change is grounds for delaying action. The US and Australia have reneged on Kyoto. During 2005 these countries, and others, suggested 'clean fuel' technologies as an alternative to emissions cuts.

In any case, according to the IPCC, the world needs to quickly improve the efficiency of its energy usage and develop renewable non-carbon fuels like: wind, solar, tidal, wave and perhaps nuclear power. It also means developing new methods of converting this clean energy into motive power, like hydrogen fuel cells for cars. Trading in Kyoto carbon permits may help.

Other less conventional solutions include ideas to stave off warming by 'mega-engineering' the planet with giant mirrors to deflect the Sun's rays, seeding the oceans with iron to generate algal blooms, or burying greenhouse gases below the sea.

The bottom line is that we will need to cut CO_2 emissions by 70% to 80% simply to stabilise atmospheric CO_2 concentrations – and thus temperatures. The quicker we do that, the less unbearably hot our future world will be.
1 September 2006

⇨ The above information is reprinted with kind permission from the *New Scientist*. Visit http://environment.newscientist.com for more information.
© *New Scientist*

World becoming more humid

Information from the Met Office

The world is becoming more humid under climate change and exacerbating global warming, research released today (11 October) reveals. Scientists from the Met Office Hadley Centre and the Climatic Research Unit at the University of East Anglia have found strong evidence of human influence on changes in global surface humidity.

In a paper published today in the scientific journal *Nature*, the scientists explore and identify the causes of changes in humidity over the last 30 years.

Surface humidity is a measure of the amount of water vapour per volume of air. Water vapour is the most important natural greenhouse gas, and its concentration is expected to rise as the climate warms due to man-made climate change, which in turn will cause more global warming.

Increases in specific humidity have implications for extreme weather, affecting:
⇨ the geographical distribution and maximum intensity of rainfall;
⇨ the maximum intensity of tropical cyclones;
⇨ heat stress;
⇨ the biosphere;
⇨ surface hydrology.

New surface humidity observational data were compared to Met Office Hadley Centre climate model output, taking into account natural and human-induced factors. This showed substantial evidence of a human influence on the changes in surface humidity, and also highlighted that natural changes alone were grossly insufficient to explain these changes.

Comparison of specific humidity with the global temperature record provides a strong degree of independent corroboration for recent – 1970s onwards – rapid warming.

Peter Thorne, climate scientist at the Met Office Hadley Centre, said: 'This confirmation, that humidity and temperature are both increasing as expected, also has important implications for future human health and comfort – especially our ability to undertake outdoor activities in a warming world.'
11 October 2007

⇨ Information from the Met Office. Visit www.metoffice.gov.uk for more information.
© *Crown copyright*

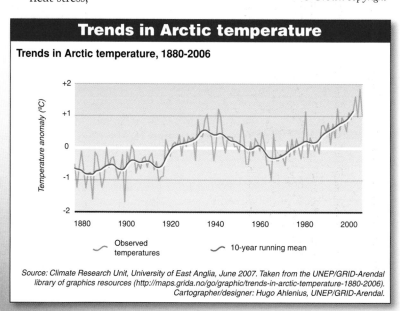

Trends in Arctic temperature

Trends in Arctic temperature, 1880-2006

Observed temperatures · 10-year running mean

Source: Climate Research Unit, University of East Anglia, June 2007. Taken from the UNEP/GRID-Arendal library of graphics resources (http://maps.grida.no/go/graphic/trends-in-arctic-temperature-1880-2006). Cartographer/designer: Hugo Ahlenius, UNEP/GRID-Arendal.

Climate change controversies

The Royal Society has produced this overview of the current state of scientific understanding of climate change to help non-experts better understand some of the debates in this complex area of science

Misleading argument 1: 'Climate change is nothing to do with humans'

It is true that the world has experienced warmer or colder periods in the past without any interference from humans. The ice ages are well-known examples of global changes to the climate. There have also been regional changes such as periods known as the 'Medieval Warm Period', when grapes were grown extensively in England, and the 'Little Ice Age', when the River Thames sometimes froze over. However, in contrast to these climate phases, the increase of three-quarters of a degree centigrade (0.74°C) in average global temperatures that we have seen over the last century is larger than can be accounted for by natural factors alone.

The Earth's climate is complex and influenced by many things – particularly changes in the Earth's orbit in relation to the Sun, which has driven the cycles of ice ages in the past, as well as volcanic eruptions and variations in the energy being emitted from the Sun. But even when we take all these factors into account, we cannot explain the temperature rises that we have seen over the last 100 years both on land and in the oceans – for example, eleven of the last twelve years have been the hottest since records started in 1850.

So what is causing this increase in average global temperature? The natural greenhouse gas effect keeps the Earth around 30°C warmer than it would otherwise be and, without it, the Earth would be extremely cold. It works because greenhouse gases such as carbon dioxide, methane, but mostly water vapour, act like a blanket around the Earth. These gases allow the Sun's rays to reach the Earth's surface but hinder the heat they create from escaping back into space. Indeed, the ability of carbon dioxide and other greenhouse gases to trap heat in this way has been understood for nearly 200 years and is regarded as firmly established science.

> **The Earth's climate is complex and influenced by many things – particularly changes in the Earth's orbit in relation to the Sun, which has driven the cycles of ice ages in the past**

Any increases in the levels of greenhouse gases in the atmosphere mean that more heat is trapped and global temperatures increase – an effect known as 'global warming'. We know from looking at gases found trapped in cores of polar ice that the levels of carbon dioxide in the atmosphere are now 35 per cent greater than they have been for at least the last 650,000 years. From the radioactivity and chemical composition of the gas we know that this is mainly due to the burning of fossil fuels, as well as the production of cement and the widespread burning of the world's forests. The increase in global temperature is consistent with what science tells us we should expect when the levels of carbon dioxide and other greenhouse gases in the atmosphere increase in the way that they have.

It has been alleged that the increased level of carbon dioxide in the atmosphere is due to emissions from volcanoes, but these account for less than one per cent of the emissions due to human activities.

Misleading argument 2: 'CO$_2$ is not responsible for global warming'

Carbon dioxide only makes up a small amount of the atmosphere, but even in tiny concentrations it has a large influence on our climate.

The properties of greenhouse gases such as carbon dioxide mean that they strongly absorb heat – a fact that can be easily demonstrated in a simple laboratory experiment. While there are larger concentrations of other gases in the atmosphere, such as nitrogen, because they do not have these heat trapping qualities they have no effect on warming the climate whatsoever.

Water vapour is the most significant greenhouse gas. It occurs naturally, although global warming caused by human activities will indirectly affect how much is in the atmosphere through, for example, increased evaporation from oceans and rivers. This will, in turn, cause either cooling or warming depending on what form such as different types of clouds the water vapour occurs in.

Humans have been adding to the effect of water vapour and other naturally occurring greenhouse gases by pumping greenhouse gases such as carbon dioxide into the atmosphere through, for example, the burning of fossil fuels and deforestation. Before industrialisation carbon dioxide made up about 0.03 per cent of the atmosphere or 280ppm (parts per million). Today, due to human influence it is about 380ppm. Even these tiny quantities have resulted in an increase in global temperatures of 0.75°C (see misleading argument 1).

Misleading argument 3: 'Rises in CO_2 occur after global warming, not before'

It is true that the fluctuations in temperatures that caused the ice ages were initiated by changes in the Earth's orbit around the Sun which, in turn, drove changes in levels of carbon dioxide in the atmosphere. This is backed up by data from ice cores which show that rises in temperature came first, and were then followed by rises in levels of carbon dioxide up to several hundred years later. The reasons for this, although not yet fully understood, are partly because the oceans emit carbon dioxide as they warm up and absorb it when they cool down and also because soil releases greenhouse gases as it

warms up. These increased levels of greenhouse gases in the atmosphere then further enhanced warming, creating a 'positive feedback'.

In contrast to this natural process, we know that the recent steep increase in the level of carbon dioxide – some 30 per cent in the last 100 years – is not the result of natural factors. This is because, by chemical analysis, we can tell that the majority of this carbon dioxide has come from the burning of fossil fuels. And, as set out in 'misleading argument 1', carbon dioxide from human sources is almost certainly responsible for most of the warming over the last 50 years. There is much evidence that backs up this explanation and none that conflicts with it.

Warming caused by greenhouse gases from human sources could lead to the release of more greenhouse gases into the atmosphere by stimulating natural processes and creating a 'positive feedback', as described above.

Misleading argument 4: 'Temperature observations don't support the theory'

It is true that in the early 1990s initial estimates of temperatures in the lowest part of the earth's atmosphere, based on measurements taken by satellites and weather balloons, did not mirror the temperature rises seen

at the Earth's surface. However, these discrepancies have been found to be related to problems with how the data was gathered and analysed and have now largely been resolved.

Our understanding of global warming leads us to expect that both the lower atmosphere – the troposphere where most greenhouse gases are found – and the surface of the Earth should warm as a result of increased levels of greenhouse gases in the atmosphere. At the same time, the lower stratosphere – the part of the atmosphere above the greenhouse gas 'blanket' – should cool.

Some have argued that climate change, as a result of human activities, isn't happening because early measurements taken from satellites and weather balloons seemed to show that virtually no warming was happening in the troposphere. However, this has been found to be due to errors in the data. Satellites were found, for example, to be slowing and dropping in orbit slightly, leading to inconsistencies in their measurements. Variations between the instruments onboard different satellites also led to discrepancies – a problem that has also been found with weather balloons. Furthermore, a mathematical error in one of the original analyses of satellite data meant that it showed less warming in the troposphere. However, once adjustments are made to take account of these and other issues, the warming in the troposphere is shown to be broadly consistent with the temperature trends we see at the Earth's surface.

In addition, the lower stratosphere has been shown to be cooling and this corresponds with our understanding of what effect global warming should have on this part of the atmosphere. However, some of this cooling is not related to increased levels of greenhouse gases but due to a different impact that humans have had on the atmosphere – the depletion of the ozone layer. Ozone warms the stratosphere by trapping incoming energy from the Sun. This reduction of ozone also has 'knock on' effects on other parts of the atmosphere, underlining the importance of taking all factors into account when looking at what is happening to our climate.

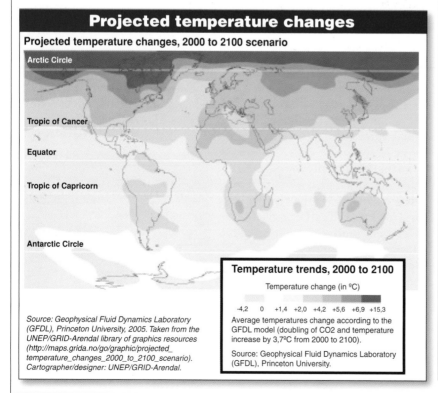

Projected temperature changes

Projected temperature changes, 2000 to 2100 scenario

Arctic Circle

Tropic of Cancer

Equator

Tropic of Capricorn

Antarctic Circle

Source: Geophysical Fluid Dynamics Laboratory (GFDL), Princeton University, 2005. Taken from the UNEP/GRID-Arendal library of graphics resources (http://maps.grida.no/go/graphic/projected_temperature_changes_2000_to_2100_scenario). Cartographer/designer: UNEP/GRID-Arendal.

Temperature trends, 2000 to 2100

Temperature change (in ºC)

-4,2   0   +1,4   +2,0   +4,2   +5,6   +6,9   +15,3

Average temperatures change according to the GFDL model (doubling of CO2 and temperature increase by 3,7ºC from 2000 to 2100).

Source: Geophysical Fluid Dynamics Laboratory (GFDL), Princeton University.

It is fair to note that in tropical regions of the world there are still some discrepancies between what computer models lead us to expect regarding temperatures at the surface and in the troposphere and what we actually see. However, these disagreements are within the bounds of the likely remaining errors in the observations and uncertainties in the models.

Misleading argument 5: 'Global warming computer models which predict the future climate are unreliable'

Modern climate models have become increasingly accurate in reproducing how the real climate 'works'. They are based on our understanding of basic scientific principles, observations of the climate and our understanding of how it functions.

By creating computer simulations of how different components of the climate system – clouds, the Sun, oceans, the living world, pollutants in the atmosphere and so on – behave and interact, scientists have been able to reproduce the overall course of the climate in the last century. Using this understanding of the climate system, scientists are then able to project what is likely to happen in the future, based on various assumptions about human activities.

It is important to note that computer models cannot exactly predict the future, since there are so many unknowns concerning what might happen. Scientists model a range of future possible climates using different scenarios of what the world will 'look like'. Each scenario makes different assumptions about important factors such as how the world's population may increase, what policies might be introduced to deal with climate change and how much carbon dioxide and other greenhouse gases humans will pump into the atmosphere. The resulting projection of the future climate for each scenario, gives various possibilities for the temperature but within a defined range.

While climate models are now able to reproduce past and present changes in the global climate rather

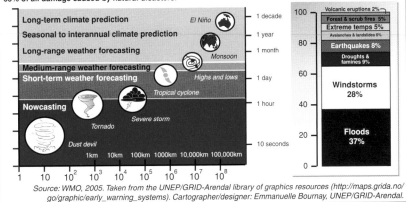

Early warning systems

Every year, disasters caused by weather, climate and water-related hazards impact on communities around the world, leading to loss of human life, destruction of social and economic infrastructure and degradation of already fragile ecosystems. Statistics from the Centre for Research on the Epidemiology of Disasters (CRED) at the University of Leuven, Belgium, reveal that from 1992-2001, about 90% of natural disasters were meteorological or hydrological in origin; the resulting economic losses were estimated at \$446bn, or about 65% of all damage caused by natural disasters.

Source: WMO, 2005. Taken from the UNEP/GRID-Arendal library of graphics resources (http://maps.grida.no/ go/graphic/early_warning_systems). Cartographer/designer: Emmanuelle Bournay, UNEP/GRID-Arendal.

well, they are not, as yet, sufficiently well-developed to project accurately all the detail of the impacts we might see at regional or local levels. They do, however, give us a reliable guide to the direction of future climate change. The reliability also continues to be improved through the use of new techniques and technologies.

Change in solar activity is one of the many factors that influence the climate but cannot, on its own, account for all the changes in global average temperature we have seen in the 20th century

---

Misleading argument 6: 'Global warming is all to do with the Sun'

Change in solar activity is one of the many factors that influence the climate but cannot, on its own, account for all the changes in global average temperature we have seen in the 20th century.

Changes in the Sun's activity influence the Earth's climate through small but significant variations in its intensity. When it is in a more 'active' phase – as indicated by a greater number of sunspots on its surface

– it emits more light and heat. While there is evidence of a link between solar activity and some of the warming in the early 20th century, measurements from satellites show that there has been very little change in underlying solar activity in the last 30 years – there is even evidence of a detectable decline – and so this cannot account for the recent rises we have seen in global temperatures.

The magnitude and pattern of changes to temperatures can only be understood by taking all of the relevant factors – both natural and human – into account. For example, major volcanic eruptions produce a cooling effect because they blast ash and other particles into the atmosphere where they persist for a few years and reduce the amount of the Sun's energy that reaches the Earth's surface. Also, burning fossil fuels produces particles called sulphate aerosols which tend to cool the climate in the same way.

Over the first part of the 20th century higher levels of solar activity combined with increases in human generated carbon dioxide to raise temperatures. Between 1940 and 1970 the carbon dioxide effect was probably offset by increasing amounts of sulphate aerosols in the atmosphere, and a slight downturn in solar activity, as well as enhanced volcanic activity.

During this period global temperatures dropped. However, in the latter part of the 20th century temperatures rose well above the levels of the 1940s. Strong measures taken to reduce sulphate pollution in some regions of the world meant that industrial aerosols began to provide less compensation for an increasing warming caused by carbon dioxide. The rising temperature during this period has been partly abated by occasional volcanic eruptions.

Misleading argument 7: 'The climate is actually affected by cosmic rays'

Any effect that cosmic rays could have on the climate is not yet very well understood but, if there is one, it is likely to be small. Cosmic rays are fast moving particles which come from space, and release electric charge in the atmosphere.

Experiments done in a laboratory hint that cosmic rays could play a role in the development of tiny particles that could in turn play a part in the formation of clouds. If this happens in the same way in the atmosphere – which isn't proven – it might lead to more clouds, which generally have a cooling effect by reflecting the Sun's rays back into space. Whether the whole chain of processes actually occurs in the atmosphere is speculative, but some of the individual steps are plausible.

It has been proposed that this process would act to enhance the influences of the Sun on the climate. We know that when the Sun is more active its magnetic field is stronger and this deflects cosmic rays away from the Earth. So the argument is that a more active Sun would lead to fewer cosmic rays reaching the Earth, resulting in fewer clouds and therefore a warmer Earth.

However, observations of clouds and galactic cosmic rays show that, at most, the possible link between cosmic rays and clouds only produces a small effect. Even if cosmic rays were shown to have a more substantial impact, the level of solar activity has changed so little over the last few decades the process could not explain the recent rises in temperature that we have seen.

Misleading argument 8: 'The negative effects of climate change are overstated'

Under one of its mid-range estimates, the Intergovernmental Panel on Climate Change (IPCC) – the world's leading authority on climate change – has projected a global average temperature increase this century of 2 to 3°C. This would mean that the Earth will experience a larger climate change than it has experienced for at least 10,000 years. The impact and pace of this change would be difficult for many people and ecosystems to adapt to.

There are real concerns that . . . rising levels of greenhouse gases in the atmosphere could set in motion large-scale and potentially abrupt changes in our planet's natural systems

---

In the short term, some parts of the world could initially benefit from climate change. For example, more northerly regions of the world may experience longer growing seasons for crops and crop yields may increase because increased levels of carbon dioxide in the atmosphere would have a fertilising effect on plants.

However, the IPCC has pointed out that as climate change progresses it is likely that negative effects would begin to dominate almost everywhere. Increasing temperatures are likely, for example, to increase the frequency and severity of weather events such as heatwaves, storms and flooding.

Furthermore there are real concerns that, in the long term, rising levels of greenhouse gases in the atmosphere could set in motion large-scale and potentially abrupt changes in our planet's natural systems and some of these could be irreversible. Increasing temperatures could, for example, lead to the melting of large ice sheets with major consequences for low-lying areas throughout the world.

And the impacts of climate change will fall disproportionately upon developing countries and the poor – those who can least afford to adapt. Thus a changing climate will exacerbate inequalities in, for example, health and access to adequate food and clean water.
April 2007

⇨ Information from the Royal Society. Visit www.royalsoc.ac.uk for more information.

© Royal Society

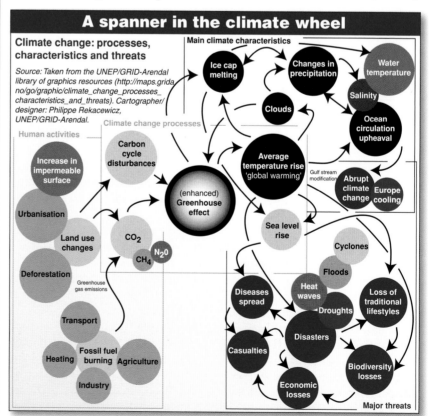

A spanner in the climate wheel

Climate change: processes, characteristics and threats

Source: Taken from the UNEP/GRID-Arendal library of graphics resources (http://maps.grida.no/go/graphic/climate_change_processes_characteristics_and_threats). Cartographer/designer: Philippe Rekacewicz, UNEP/GRID-Arendal.

Why it's green to go vegetarian

Information from the Vegetarian Society

However much we might like to believe the sceptics, there is a very broad scientific consensus that our climate is changing and mankind is, at least in part, responsible.

'Greenhouse gases' are so called because they act like the glass of a greenhouse, trapping heat from the sun to warm up the Earth. Most of these gases occur naturally and without them our planet would be too cold to sustain life, but the balance is a very delicate one. Modern humans are causing a massive increase in greenhouse gas emissions and with too much of these gases in the atmosphere, temperatures will rise higher and higher.

At the beginning of 2007, the United Nation's Intergovernmental Panel on Climate Change (IPCC) reported that global temperatures will probably rise by between 1.8 and 4°C by the end of this century (the possible range being between 1.1 to 6.4°C). This may not sound like a lot but the polar ice caps are already melting and the report predicted that these temperature changes would cause rises in sea levels and increases in the number of hurricanes and tropical storms. When the sea level rises, low-lying land around the world is threatened and over time, things just get worse as the expanding oceans increase further, thanks to the melting of ice sheets covering Greenland and Antarctica.

Many scientists and world leaders believe that climate change is the most serious issue facing the whole human race.

The most important greenhouse gases are carbon dioxide (CO_2), methane (CH_4) and nitrous oxide (N_2O). The atmospheric concentrations of all three have increased phenomenally in modern times. Comparing figures from 2005 with pre-industrialised

levels (measurements from 1750), carbon dioxide has increased from around 280 parts per million (ppm) to 379ppm, methane has increased from 715 parts per billion (ppb) to 1774ppb and nitrous oxide has increased from 270 ppb to 319 ppb. The increase in carbon dioxide is due mostly to the use of fossil fuels and changes in the way we use land.

Increases of methane and nitrous oxide, however, are primarily caused by agriculture.

Farmed animals produce more greenhouse gas emissions (18%) than the world's entire transport system (13.5%). Cows' flatulence, alongside animal excrement, makes the headlines due to both of them being extremely damaging. However, farming animals also generates gaseous emissions through the manufacture of fertilisers (to grow feed crops), industrial feed production and the transportation of both live animals and their carcasses across the globe.

9% of human-related CO_2 emissions are caused by the livestock sector, mostly due to changes in land use (e.g. forests being cleared for grazing or growing animal feed) and the use of fossil fuels for farm operations.

Methane has 23 times the global warming impact of CO_2 and ruminant mammals (cows and sheep) are responsible for 37% of the total methane generated by human activity.

There are approximately 1.5 billion cattle and 1.7 billion sheep on the planet. A single cow can produce as much as 500 litres of methane per day.

Nitrous oxide is almost 300 times as damaging to the climate as carbon dioxide with 65% of the total quantity produced by human activity coming from livestock (mostly their manure).

The animals we rear for meat also account for 64% of all the ammonia we humans impose on our precious atmosphere, contributing significantly to acid rain.

⇨ The above information is reprinted with kind permission from the Vegetarian Society. Visit www.vegsoc.org for more information on this and related issues.

© *Vegetarian Society*

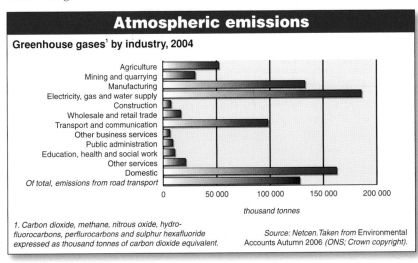

Atmospheric emissions

Greenhouse gases¹ by industry, 2004

thousand tonnes

1. Carbon dioxide, methane, nitrous oxide, hydro-fluorocarbons, perflurocarbons and sulphur hexafluoride expressed as thousand tonnes of carbon dioxide equivalent.

Source: Netcen. Taken from Environmental Accounts Autumn 2006 *(ONS; Crown copyright).*

What does climate change mean for us?

Information from Stop Climate Chaos

Feeling hot, hot, hot

The world is warming faster than at any time in the last 10,000 years. The global average temperature will increase between 1.4°C to 5.8°C by 2100. An increase of 2°C would massively impact on coral reefs, arctic systems and local communities.

Stop Climate Chaos believes that temperature rise must stay below 2°C in order to limit dangerous climate change.

150,000 people already die every year from climate change

Here at home 10 August 2003 was the hottest day so far recorded in Britain: the highest temperature was 38.1°C (more than 100°F). And as cloud cover decreases, there will be increased exposure to harmful ultra-violet rays, which cause skin cancer.

The European heatwave in August 2003 – linked directly by many scientists to climate change – was the hottest in 500 years and killed 28,000 people. The likelihood of such heatwaves may triple by the 2080s as a result of climate change. Cities such as Athens, Delhi and Chicago have sweltered under heatwaves and seen death tolls rise.

Harvest for the world...?

The World Health Organisation say that 150,000 people already die every year from climate change. And people in developing countries are four times more likely to die in natural disasters than people in developed countries.

Poor people – especially children – are the least able to adapt to changes caused by climate change. They are the most vulnerable to 'natural' disasters; the least able to

move from affected regions; and the most reliant on harvests coming at the right time.

Oh I do like to be beside the seaside

Global sea level has risen globally between 0.1 and 0.2 metres during the 20th century and could rise by almost a metre by 2100. 100 million more people will be flooded by end of century.

Sea-level rise will help result in about 40-50 per cent of the world's coastal wetlands being lost by the 2080s. Rising sea levels threaten entire nations on low-lying islands in the Pacific and Indian Oceans; the inhabitants of Tuvalu have had to be evacuated and sea-level rise is causing coastal erosion in the Maldives.

If the West Antarctic ice shelf breaks away further then climate change global sea levels could rise by 5 metres or more by 2100. This would swamp many of the world's major cities including New York and Shanghai. In the UK, Hull, Cardiff, Portsmouth – even London – could be under water.

In the Arctic too, the dramatic impacts of climate change are already being seen. Glaciers and ice caps are continuing their widespread retreat during the 21st century and, by 2080, Arctic sea ice could completely disappear during the summer months.

Stormy weather

Scientists predict that hurricanes and tornadoes will increase in intensity and range as a result of climate change. This means that category 4 and 5 storms – like the one that flooded New Orleans – will become more and more common.

Exodus

Climate change could spark regional conflicts as millions of environmental refugees flee from floods and droughts, and food and water shortages. This figure could reach 150 million by the end of the century.

Money, money, money

The economic costs of global warming are doubling every decade and the insurance industry puts the financial cost due to climate change at hundreds of billions of dollars each year. Just one example – the total cost to the global economy of losing half the world's coral reefs has been estimated at $140 billion.

⇨ The above information is reprinted with kind permission from Stop Climate Chaos, a coalition which brings together over 60 organisations from environment and development charities to unions, faith and women's groups. Visit http://icount.org.uk for more.

© Stop Climate Chaos

ONLY TWO LEFT!
EXCITING LAND RELEASE!
Well above rising sea levels
Beat the rush, buy now!
Everest developments

Biodiversity and climate change: ecosystems

Information from the United Nations Environment Programme World Conservation Monitoring Centre

Climate change will increasingly drive biodiversity loss, affecting both individual species and their ecosystems. An ecosystem can be defined as a community of plant and animal species and the physical environment that they occupy, which includes the climate regime. When climate conditions change, unexpected results may follow. Each species will respond in an individual fashion, according to its climate tolerances and its ability to disperse into a new location, alter its phenology (e.g. breeding date) or adapt to shifting food sources. It is difficult to predict the overall result of changes in the abundance of herbivores and food plants, predators and prey.

Many studies have attempted to project the rate and extent of terrestrial ecosystem response to climate change, some using simple models assuming that entire ecosystems will shift to follow the changing climate, and others using 'plant functional type' models featuring the responses of different types of herbs, bushes and trees. Vegetation zones are typically expected to move towards higher latitudes or higher altitudes following shifts in average temperatures.

The vulnerability of an ecosystem to climate change depends on its species' tolerance of change, the degree of change, and the other stresses already affecting it. For example, coral reefs already polluted by sediment and nutrient run-off may find it more difficult to survive increasing ocean temperatures. Climate change can also increase an ecosystem's vulnerability to existing pressures. For example, where fire is used to clear agricultural land, drier, warmer conditions will make an adjacent forest more susceptible to burning. In addition, disturbances such as fires, floods and insect plagues are expected to become more frequent as a result of climate change.

Up to a point, the increased concentrations of atmospheric carbon dioxide that are driving global warming also have a direct effect on plants, both increasing rates of photosynthesis and improving water use efficiency. This increases tolerance to drought, so may help some terrestrial ecosystems to withstand the effects of climate change.

Marine ecosystems will be affected not only by an increase in sea temperature and changes in ocean circulation, but also by ocean acidification, as the concentration of dissolved carbon dioxide (carbonic acid) rises. This is expected to negatively affect shell-forming organisms, corals and their dependent ecosystems, with some researchers warning of catastrophic results.

Polar ecosystems are especially vulnerable to climate change, with effects such as thawing permafrost, decreased snow cover, losses from ice sheets and changes in ocean temperatures. Impacts on Arctic biodiversity are already being observed.

Resources

⇨ *Climate Change 2007: Impacts, Adaptation and Vulnerability.* IPCC Summary for Policymakers (2007)

⇨ A world without corals? *Science* (2007)

⇨ *Our Precious Coasts* (2006)

⇨ A global overview of the conservation status of tropical dry forests. *Journal of Biogeography* (2006)

⇨ *Climate change and biodiversity.* Yale University Press (2006)

⇨ *Global Climate Change and Biodiversity* (2003)

⇨ *Climate change and biodiversity.* IPCC (2001)

⇨ The above information is reprinted with kind permission from the United Nations Environment Programme World Conservation Monitoring Centre. Visit www2.wcmc.org.uk for more information.

© *United Nations Environment Programme World Conservation Monitoring Centre*

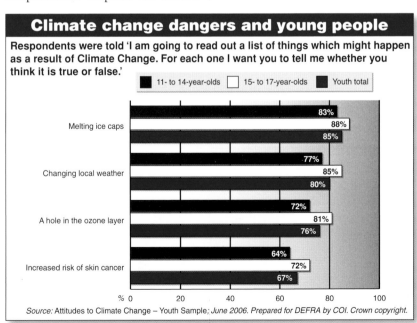

Climate change dangers and young people

Respondents were told 'I am going to read out a list of things which might happen as a result of Climate Change. For each one I want you to tell me whether you think it is true or false.'

■ 11- to 14-year-olds ☐ 15- to 17-year-olds ■ Youth total

Melting ice caps: 83%, 88%, 85%

Changing local weather: 77%, 85%, 80%

A hole in the ozone layer: 72%, 81%, 76%

Increased risk of skin cancer: 64%, 72%, 67%

% 0   20   40   60   80   100

Source: Attitudes to Climate Change – Youth Sample; *June 2006. Prepared for DEFRA by COI. Crown copyright.*

Climate change and cities

Information from the International Institute for Environment and Development

What is done, or not done, in cities in relation to climate change over the next 5-10 years will affect hundreds of millions of people, because their lives and livelihoods are at risk from global warming. What is done in cities will also have a major influence on whether the escalating risks for the whole planet will be reduced or eliminated. Climate change needs to be considered in all development plans and investments – local, regional, national and international. Urban growth must be made more climate-resilient and help reduce, rather than increase, greenhouse gas emissions. This will not be done by the market; it can only be done by governments.

Innovation and action on climate change needs to focus on urban centres in Africa, Asia and Latin America. Certainly, urban areas in high-income nations are currently the greatest threat, because of their historic and current contributions to greenhouse gas emissions. Current levels of greenhouse gas emissions per person are particularly high in cities in North America and Australia – often 25 to 50 times higher than in cities in low-income nations. But there are three reasons why climate change action is urgently needed in urban centres in low- and middle-income nations:

⇨ They already have three-quarters of the world's urban population; China's urban population is as large as Europe's; India and Africa both have larger urban populations than North America;

⇨ They will house most of the growth in the world's population over the next ten to twenty years, and have a major impact on future greenhouse gas emissions;

⇨ They have a large and growing proportion of the world's population most at risk from storms,

By David Satterthwaite

floods and other climate change-related impacts.

For cities, perhaps the most obvious increased risk from climate change comes from the increased number and intensity of extreme weather events, such as heavy rainstorms, cyclones and hurricanes. Cities on the coast are obviously more at risk from sea-level rise, but perhaps the greatest risk comes from a combination of storm surges and high tides. Rising sea levels may also mean rising water tables that undermine building foundations and increased saline water intrusion into valuable groundwater sources. Many non-coastal cities face serious problems with flooding – as they are beside rivers or in foothills of mountains and so vulnerable to more intense precipitation or snowmelt.

There are many less dramatic but nonetheless serious risks from climate change, especially for low-income groups. Many cities will get less rainfall, and most will experience more heatwaves and greater air pollution problems. Many city

Key messages

⇨ Urban centres in low- and middle-income nations provide homes to a large population increasingly vulnerable to the direct and indirect impacts of climate change.

⇨ Most cities are at risk from extreme weather events independent of climate change because of the lack of investment in basic infrastructure.

⇨ Adaptation can fit within development plans and existing institutions. Investing in measures to help adaptation to climate change will usually reduce this risk as well. Many measures that are pro-poor also reduce risks from climate change. Adaptation and mitigation measures incorporated in the way cities develop can bring large benefits at relatively low cost. Action should be taken as infrastructure expands, rather than when already in place.

⇨ Good urban governance supported by a good national framework is key to moving quickly, efficiently and at scale.

economies will suffer from decreasing possibilities for agriculture in their surroundings. Tourist cities on the coast will have their assets compromised by flood damage to coastal reefs and loss of beaches. Warmer average temperatures will expand the areas in which tropical diseases can occur. And while some changes may provide positive opportunities, these will require adaptation.

The quality of government – at national and city level – has a very large influence on how risks from climate change are dealt with, especially for people with limited incomes or assets. For instance: the quality of infrastructure to limit risks of flooding for the whole city area and not just for wealthier areas; the level of disaster-preparedness, including warnings of approaching storms and floods; and the ability to offer disaster-response to help those who have lost their homes and livelihoods.

There is a profound unfairness globally – those most at risk from climate change are those least responsible for most greenhouse gas emissions. The very survival of some islands and low-income nations is in doubt as much of their land area is at risk from sea-level rise, yet their contributions to global greenhouse emissions have been very small. There are also tens of millions of people in Asia and Africa whose homes and livelihoods are at risk from sea-level rise and storms yet they have made very little contribution to greenhouse gas emissions. Most global warming to date has been driven by consumers in high-income nations and the production systems that serve them, with the USA being the largest contributor, both historically and currently. Would the US Government oppose the Kyoto Protocol's modest targets for emission reductions if Washington DC, New York and Los Angeles faced risks comparable to those facing Dhaka, Mumbai and Bangkok today?

If action is taken now, there are large cumulative benefits and large cost savings, including avoidance of premature death, injury and property loss. The earlier action is taken to reduce greenhouse gas emissions and to begin reducing vulnerability to climate change, the lower the costs. Urban centres need a planning and investment framework that breaks the link between growing incomes and rising greenhouse gas emissions per person. Evidence shows how this can be done: better-designed housing and office buildings, with much less need for heating or cooling and artificial light; a city in which all income groups prefer to make

most journeys by walking, bicycling or public transport; cities where industry, commerce and services demonstrate their capacity to cut energy requirements and reduce waste.

There is a profound unfairness globally – those most at risk from climate change are those least responsible for most greenhouse gas emissions

---

If initiated now, action on climate change, to reduce greenhouse gas emissions and risks from its effects, need not draw resources from other pressing tasks. There are obvious worries that such action will draw resources from other priorities. Most cities in Africa and Asia and many in Latin America have 33-50% of populations lacking good provision for water and sanitation and living in illegal settlements, because they cannot afford the cost of legal housing. Some 900 million urban inhabitants live in very poor quality shelter – often with two to three persons per room. Most urban centres have a large backlog to make up in basic infrastructure. So it is difficult to see action on climate change as a priority. But there are three good reasons for taking action now:

1. There are many ways to reduce greenhouse gas emissions with modest adjustments to investment by choosing low carbon technologies which, over time, produce much lower levels of greenhouse gas emissions, even in cities with booming economies;

2. Much of what needs to be done to reduce risks from climate change also reduces other risks; for instance, better drainage systems also protect health and reduce risks of flooding and water-logging;

3. Part of what needs to be done does not require additional government expenditure but is achieved by changing regulatory frameworks that influence individual, household, community, company and corporate investments – for instance adjustments to building regulations, land use plans, land subdivision regulations, pollution control and waste management.

These adjustments will not be easy since many will face opposition from powerful vested interests. In addition, changing regulations on buildings, infrastructure and land subdivision needs to avoid imposing added costs on low-income households.

Investments to counter climate change and protect cities from its effects must avoid being anti-poor in their impacts. For instance, it is common for low-income areas, and many informal settlements, to be in areas that are poorly drained or most at risk from flooding. If climate change-related policies are also to benefit the inhabitants of these areas this means fully involving them in plans to reduce flooding and other risks. Relocation should be avoided wherever possible – and

Human activity and climate change

Respondents were told 'I am going to read out a number of statements about climate change. For each one, I would like you to tell me if you agree or disagree.'

■ Strongly agree	□ Tend to agree	■ Neither agree nor disagree	▨ Tend to disagree	▨ Strongly disagree	■ Undecided/ don't know

	Strongly agree	Tend to agree	Neither agree nor disagree	Tend to disagree	Strongly disagree	Undecided/don't know
Human activity does not have significant effect on the climate	4	14%	9%	31%	38%	5
Climate change is too complex and uncertain for scientists to make useful forecasts	8%	32%	15%	25%	13%	7%
Many leading experts still question if human activity is contributing to climate change	11%	45%	13%	14%	8%	9%
Nothing I can do personally could avert climate change	7%	19%	8%	32%	30%	5
Ultimately, I am confident that the world community can find a solution to the problems posed by climate change	8%	38%	15%	21%	11%	7%

% 0   20   40   60   80   100

Source: Ipsos MORI, 4 July 2007. Ipsos MORI interviewed a representative sample of 2,031 adults aged 16+ face-to-face, in home, on 14-20 June 2007. Data are weighted to known population profile.

upgrading programmes favoured, in which governments work with the inhabitants of these areas to combine improved infrastructure (for instance for water, sanitation, drainage, support for house improvements) with much lower flooding risks. Low-income groups may be prepared to move from hazardous sites, but only if they are involved in decisions about where to move and how the move is organised. This means a fundamental change in practice, since most city governments tend to move those people who live in the path of new infrastructure investments, and push them to peripheral areas, destroying their homes, asset base, social networks, and their incomes.

Well-governed cities show how to de-link development from high greenhouse gas emissions. The concentration of people and production in cities facilitates many investments and actions that keep down energy requirements for buildings, transport systems and enterprises and that also support waste reduction. Much valuable work can be done to exchange and link lessons of good practice.

'Climate change-aware' development policies can contribute much to lowering greenhouse gas emissions and to producing city economies and populations that are less at risk. But too many policy-makers at national and city levels see climate change as an environmental issue or a global issue that is not their concern. Too many climate change specialists have little knowledge of development, as their approach focuses on reducing greenhouse emissions alone, and not on helping nations and cities learn how to change and adapt. Equally, climate change science deals mainly with global and regional impacts and is less able to provide reliable assessments for city regions.

⇨ The above information is reprinted with kind permission from the International Institute for Environment and Development (IIED). Please visit www.iied.org for more information.

© IIED 2006

Climate change denial

The threat is from those who accept climate change, not those who deny it

If the biosphere is ruined it will be done by people who know that emissions must be cut – but refuse to alter the way they live, says George Monbiot.

You have to pinch yourself. Until now the *Sun* has denounced environmentalists as 'loonies' and 'eco beards'. Last week it published 'photographic proof that climate change is real'. In a page that could have come straight from a Greenpeace pamphlet, it laid down 10 'rules' for its readers to follow: 'Use public transport when possible; use energy-saving light bulbs; turn off electric gadgets at the wall; do not use a tumble dryer ...'

Two weeks ago the *Economist* also recanted. In the past it has asserted that 'Mr Bush was right to reject the prohibitively expensive Kyoto pact'.

It co-published the Copenhagen Consensus papers, which put climate change at the bottom of the list of global priorities. Now, in a special issue devoted to scaring the living daylights out of its readers, it maintains that 'the slice of global output that would have to be spent to control emissions is probably ... below 1%'. It calls for carbon taxes and an ambitious programme of government spending.

Almost everywhere, climate change denial now looks as stupid and as unacceptable as Holocaust denial. But I'm not celebrating yet. The danger is not that we will stop talking about climate change, or recognising that it presents an existential threat to humankind. The danger is that we will talk ourselves to kingdom come.

If the biosphere is wrecked, it will not be done by those who couldn't give a damn about it, as they now belong to a diminishing minority. It will be destroyed by nice, well-meaning, cosmopolitan people who accept the case for cutting emissions, but who won't change by one iota the way they live. I know people who profess to care deeply about global warming, but who would sooner drink Toilet Duck than get rid of their Agas, patio heaters and plasma TVs, all of which are staggeringly wasteful. A recent brochure published by the Co-operative Bank boasts that its 'solar tower' in Manchester 'will generate enough electricity every year to make 9 million cups of tea'. On the previous page it urges its customers 'to live the dream and purchase that perfect holiday home ... With low cost flights now available, jetting off to your home in the sun at the drop of a hat is far more achievable than you think.'

Environmentalism has always been characterised as a middle-class concern; while this has often been unfair, there is now an undeniable nexus of class politics and morally superior consumerism. People allow themselves to believe that their impact on the planet is lower than that of the great unwashed because they shop at Waitrose rather than Asda, buy Tomme de Savoie instead of processed cheese slices and take eco-safaris in the Serengeti instead of package holidays in Torremolinos. In reality, carbon emissions are closely related to income: the richer you are, the more likely you are to be wrecking the planet, however much stripped wood and hand-thrown crockery there is in your kitchen.

It doesn't help that pol-iticians, businesses and even climate-change campaigners seek to shield us from the brutal truth of just how much has to change. Last week Friends of the Earth published the report it had commissioned from the Tyndall Centre for Climate Change Research, which laid out the case for a 90% reduction in carbon emissions by 2050. This caused astonishment in the media. But other calculations, using the same sources, show that even this ambitious target is two decades too late. It becomes rather complicated, but please bear with me, for our future rests on these numbers.

The Tyndall Centre says that to prevent the earth from warming by more than two degrees above preindustrial levels, carbon dioxide concentrations in the atmosphere must be stabilised at 450 parts per million or less (they currently stand at 380). But this, as its sources show, is plainly insufficient. The reason is that carbon dioxide (CO_2) is not the only greenhouse gas. The others – such as methane, nitrous oxide and hydrofluorocarbons – boost its

impacts by around 15%. When you add the concentrations of CO_2 and the other greenhouse gases together, you get a figure known as 'CO_2 equivalent'. But the Tyndall Centre uses 'CO_2' and 'CO_2 equivalent' interchangeably, permitting an embarrassing scientific mish-mash.

'Concentrations of 450 parts per million CO_2 equivalent or lower', it says, provide a 'reasonable to high probability of not exceeding 2ºC'. This is true, but the report is not calling for a limit of 450 parts of 'CO_2 equivalent'. It is calling for a limit of 450 parts of CO_2, which means at least 500 parts of CO_2 equivalent. At this level there is a low to very low probability of keeping the temperature rise below two degrees. So why on earth has this reputable scientific institution muddled the figures?

You can find the answer on page 16 of the report. 'As with all client-consultant relationships, boundary conditions were established within which to conduct the analysis ... Friends of the Earth, in conjunction with a consortium of NGOs and with increasing cross-party support from MPs, have been lobbying hard for the introduction of a "climate change bill" ... [The bill] is founded essentially on a correlation of 2ºC with 450 parts per million of CO_2.'

In other words, Friends of the Earth had already set the target before it asked its researchers to find out

what the target should be. I suspect that it chose the wrong number because it believed a 90% cut by 2030 would not be politically acceptable.

This echoes the refusal of Sir David King, the government's chief scientist, to call for a target of less than 550 parts per million of CO_2 in the atmosphere, on the grounds that it would be 'politically unrealistic'. The message seems to be that the science can go to hell – we will tell people what we think they can bear.

So we all deceive ourselves and deceive each other about the change that needs to take place. The middle classes think they have gone green because they buy organic cotton pyjamas and handmade soaps with bits of leaf in them – though they still heat their conservatories and retain their holiday homes in Croatia. The people who should be confronting them with hard truths balk at the scale of the challenge. And the politicians won't jump until the rest of us do.

On Sunday the Liberal Democrats announced that they are making climate change their top political priority, and on Tuesday they voted to shift taxation from people to pollution. At first sight it looks bold, but then you discover that they have scarcely touched the problem. While total tax receipts in the United Kingdom amount to £350bn a year, they intend to shift just £8bn – or 2.3%.

So the question which now confronts everyone – politicians, campaign groups, scientists, readers of the *Guardian* as well as the *Economist* and the *Sun* – is this: how much reality can you take? Do you really want to stop climate chaos, or do you just want to feel better about yourself?

⇨ George Monbiot's book *Heat: How to Stop the Planet Burning* is published by Allen Lane next week. He has also launched a website – turnuptheheat.org – exposing false environmental claims made by corporations and celebrities.

First appeared on the Guardian, *21 September 2006*

© *George Monbiot*

The deceit behind global warming

By Christopher Booker and Richard North

No one can deny that in recent years the need to 'save the planet' from global warming has become one of the most pervasive issues of our time. As Tony Blair's chief scientific adviser, Sir David King, claimed in 2004, it poses 'a far greater threat to the world than international terrorism', warning that by the end of this century the only habitable continent left will be Antarctica.

Inevitably, many people have been bemused by this somewhat one-sided debate, imagining that if so many experts are agreed, then there must be something in it. But if we set the story of how this fear was promoted in the context of other scares before it, the parallels which emerge might leave any honest believer in global warming feeling uncomfortable.

The story of how the panic over climate change was pushed to the top of the international agenda falls into five main stages. Stage one came in the 1970s when many scientists expressed alarm over what they saw as a disastrous change in the earth's climate. Their fear was not of warming but global cooling, of 'a new Ice Age'.

For three decades, after a sharp rise in the interwar years up to 1940, global temperatures had been falling. The one thing certain about climate is that it is always changing. Since we began to emerge from the last Ice Age 20,000 years ago, temperatures have been through significant swings several times. The hottest period occurred around 8,000 years ago and was followed by a long cooling. Then came what is known as the 'Roman Warming', coinciding with the Roman empire. Three centuries of cooling in the Dark Ages were followed by the 'Mediaeval Warming', when the evidence agrees the world was hotter than today.

Around 1300 began 'the Little Ice Age', that did not end until 200 years ago, when we entered what is known as the 'Modern Warming'. But even this has been chequered by colder periods, such as the 'Little Cooling' between 1940 and 1975. Then, in the late 1970s, the world began warming again.

> ### No one can deny that in recent years the need to 'save the planet' from global warming has become one of the most pervasive issues of our time

A scare is often set off – as we show in our book with other examples – when two things are observed together and scientists suggest one must have been caused by the other. In this case, thanks to readings commissioned by Dr Roger Revelle, a distinguished American oceanographer, it was observed that since the late 1950s levels of carbon dioxide in the earth's atmosphere had been rising. Perhaps it was this increase that was causing the new warming in the 1980s?

Stage two of the story began in 1988 when, with remarkable speed, the global warming story was elevated into a ruling orthodoxy, partly due to hearings in Washington chaired by a youngish senator, Al Gore, who had studied under Dr Revelle in the 1960s.

But more importantly global warming hit centre stage because in 1988 the UN set up its Inter-governmental Panel on Climate Change (the IPCC). Through a series of reports, the IPCC was to advance its cause in a rather unusual fashion. First it would commission as many as 1,500 experts to produce a huge scientific report, which might include all sorts of doubts and reservations. But this was to be prefaced by a Summary for Policymakers, drafted in consultation with governments and officials – essentially a political document – in which most of the caveats contained in the experts' report would not appear.

This contradiction was obvious in the first report in 1991, which led to the Rio conference on climate change in 1992. The second report in 1996 gave particular prominence to a study by an obscure US government scientist claiming that the evidence for a connection between global warming and rising CO_2 levels was now firmly established. This study came under heavy fire from various leading climate experts for the way it manipulated the evidence. But this was not allowed to stand in the way of the claim that there was now complete scientific consensus behind the CO_2 thesis, and the Summary for Policymakers, heavily influenced from behind the scenes by Al Gore, by this time US Vice-President, paved the way in 1997 for the famous Kyoto Protocol.

Kyoto initiated stage three of the story, by formally committing governments to drastic reductions in their CO_2 emissions. But the treaty still had to be ratified and this seemed a good way off, not least thanks to its rejection in 1997 by the US Senate, despite the best attempts of Mr Gore.

Not the least of his efforts was his bid to suppress an article co-authored by Dr Revelle just before his death. Gore didn't want it to be known that his guru had urged that the global warming thesis should be viewed with more caution.

One of the greatest problems Gore and his allies faced at this time was the mass of evidence showing that in the past, global temperatures had been higher than in the late 20th century.

In 1998 came the answer they were looking for: a new temperature chart, devised by a young American physicist, Michael Mann. This became known as the 'hockey stick' because it showed historic temperatures running in an almost flat line over the past 1,000 years, then suddenly flicking up at the end to record levels.

If global warming does turn out to have been a scare like all the others, it will certainly represent as great a collective flight from reality as history has ever recorded

---

Mann's hockey stick was just what the IPCC wanted. When its 2001 report came out it was given pride of place at the top of page 1. The Mediaeval Warming, the Little Ice Age, the 20th century Little Cooling, when CO_2 had already been rising, all had been wiped away.

But then a growing number of academics began to raise doubts about Mann and his graph. This culminated in 2003 with a devastating study by two Canadians showing how Mann had not only ignored most of the evidence before him but had used an algorithm that would produce a hockey stick graph whatever evidence was fed into the computer. When this was removed, the graph re-emerged just as it had looked before, showing the Middle Ages as hotter than today.

It is hard to recall any scientific thesis ever being so comprehensively discredited as the 'hockey stick'. Yet the global warming juggernaut rolled on regardless, now led by the European Union. In 2004, thanks to a highly dubious deal between the EU and Putin's Russia, stage four of the story began when the Kyoto treaty was finally ratified.

In the past three years, we have seen the EU announcing every kind of measure geared to fighting climate change, from building ever more highly-subsidised wind turbines, to a commitment that by 2050 it will have reduced carbon emissions by 60 per cent. This is a pledge that could only be met by such a massive reduction in living standards that it is impossible to see the peoples of Europe accepting it.

All this frenzy has rested on the assumption that global temperatures will continue to rise in tandem with CO_2 and that, unless mankind takes drastic action, our planet is faced with the apocalypse so vividly described by Al Gore in his Oscar-winning film *An Inconvenient Truth*.

Yet recently, stage five of the story has seen all sorts of question marks being raised over Gore's alleged consensus. For instance, he claimed that by the end of this century world sea levels will have risen by 20ft when even the IPCC in its latest report, only predicts a rise of between four and 17 inches. There is also of course the harsh reality that, wholly unaffected by Kyoto, the economies of China and India are now expanding at nearly 10 per cent a year, with China likely to be emitting more CO_2 than the US within two years.

More serious, however, has been all the evidence accumulating to show that, despite the continuing rise in CO_2 levels, global temperatures in the years since 1998 have no longer been rising and may soon even be falling.

It was a telling moment when, in August, Gore's closest scientific ally, James Hansen of the Goddard Institute for Space Studies, was forced to revise his influential record of US surface temperatures showing that the past decade has seen the hottest years on record. His graph now concedes that the hottest year of the 20th century was not 1998 but 1934, and that four of the 10 warmest years in the past 100 were in the 1930s.

Furthermore, scientists and academics have recently been queuing up to point out that fluctuations in global temperatures correlate more consistently with patterns of radiation from the sun than with any rise in CO_2 levels, and that after a century of high solar activity, the sun's effect is now weakening, presaging a likely drop in temperatures.

If global warming does turn out to have been a scare like all the others, it will certainly represent as great a collective flight from reality as history has ever recorded. The evidence of the next 10 years will be very interesting.

⇨ *Scared to Death: From BSE To Global Warming – How Scares Are Costing Us The Earth* by Christopher Booker and Richard North (Continuum, £16.99) is available for £14.99 + £1.25 p&p. To order call Telegraph Books on 0870 428 4115 or go to books.telegraph.co.uk
4 November 2007

G8 climate change accord elicits mixed reactions

Information from OneWorld.net

By Ramesh Jaura

The climate change accord clinched at the G8 summit in the seaside resort of Heiligendamm has given rise to satisfaction and scepticism among experts.

The agreement Thursday (7 June) has 'paved the way for negotiations in Bali in December and given climate talks under the auspices of the UN a considerable boost,' said Yvo de Boer, the Executive Secretary of the United Nations Framework Convention on Climate Change (UNFCCC).

'The multilateral climate change process under the United Nations has been re-energised,' he said. 'This is a breakthrough in terms of making progress towards an enhanced future climate change regime and will send important signals to developing countries on the readiness of industrialised nations and emerging economies to act,' he added.

Alluding to the large emerging economies of China, India, Brazil, South Africa and Mexico, the UN's top climate change official said: 'There is now a need to engage these economies on how best to address the challenges of climate change. It is very encouraging that the G8 is ready to work with the +5 countries on long-term strategies and that major emitters of the process will report back to the UNFCCC by 2008.'

According to the G8 communiqué, negotiations under the UNFCCC should be finished by 2009. This would give governments enough time to ratify the agreement before the first commitment period of the Kyoto Protocol expires in 2012.

The G8 statement said: 'We are committed to moving forward in that forum and call on all parties to actively and constructively participate in the UN Climate Change Conference in Indonesia in December 2007 with a view to achieving a comprehensive post 2012-agreement (post Kyoto-agreement) that should include all major emitters.'

But Friends of the Earth International Climate Change campaigner Yuri Onodera said: 'We have already seen many empty promises by G8 leaders over the past years but there has not been much real action, so we urge G8 leaders to act now and cut their greenhouse gas emissions drastically and immediately.'

> **Collectively the G8, which represent just 13 per cent of the world's population, are responsible for around 43 per cent of the world's greenhouse gas emissions**

G8 nations had so far failed to take their historical responsibilities seriously and pay the ecological debt they owe to the people in poorer countries who are suffering from the consequences of the current unsustainable development model, Onodera said.

But he welcomed that 'the US administration, which continuously obstructed the fight against climate change, did not manage to prevent world leaders here from pledging that they will take multilateral action.'

Collectively the G8 – Britain, France, Germany, Italy, Canada, USA, Japan and Russia – which represent just 13 per cent of the world's population, are responsible for around 43 per cent of the world's greenhouse gas emissions.

All countries except the US and Russia made a non-binding pledge to cut the climate change-causing gases by at least half by 2050. Scientists say that such a cut is necessary to try and keep the increase in global average temperatures below two degrees centigrade from pre-industrial levels.

Agreeing with UNFCCC's Executive Secretary, Germanwatch NGO's political director Christoph Bals said the G8 accord had 'pushed open the door to serious UN climate change negotiations'. He pointed out that all G8 nations had agreed to take note of and expressed concern about the recent UN Intergovernmental Panel on Climate Change (IPCC) reports.

The declaration said: 'The most recent report concluded both, that global temperatures are rising, that this is caused largely by human activities and, in addition, that for increases in global average temperature, there are projected to be major changes in ecosystem structure and function with predominantly negative consequences for bio-diversity and ecosystems, e.g. water and food supply.'

Bals said the G8 had also recognised the need for legally binding UN accord that envisages substantial greenhouse gas reductions. The statement said: 'Taking into account the scientific knowledge as represented in the recent IPCC reports, global greenhouse gas emissions must stop rising, followed by substantial global emission reductions.'

The Germanwatch said: 'In setting a global goal for emissions reductions in the process we have agreed today involving all major emitters, we will consider seriously the decisions made by the European Union, Canada and Japan which include at least a halving of global emissions by 2050.'

UNFCCC's Executive Secretary de Boer said: 'It will now be critical to have everything in place so that the negotiation process can be set in motion at the United Nations Climate Change Conference in Bali in December of this year.'

One key focus of the document is on adaptation, with G8 leaders acknowledging that considerable funds will be needed to above all enable the most vulnerable to adapt to the inevitable effects of climate change and expressing a willingness to work with developing countries on the issues.

Another key element of the document, according to the UNFCCC, is the call to expand the Kyoto Protocol's Clean Development Mechanism (CDM). The CDM permits industrialised countries to invest in sustainable development projects in developing countries, and thereby generate tradable emission credits.

The Kyoto Protocol presently requires 36 industrialised countries and the European Community to reduce greenhouse gas emissions by an average of 5 per cent below 1990 levels in its first commitment period between 2008 and 2012.

The CDM is currently undergoing a boom and is expected to generate around two billion in certified emission reductions (CERs) by 2012. One CER amounts to one tonne of CO_2 equivalent.

According to de Boer, such mechanisms need to be part of any meaningful post-2012 climate change regime. 'If half of the emission reductions would be met through investments in developing countries, for example via the CDM, there is a potential to generate up to 100 billion dollars per year in green investment flows to developing countries.'

'The door has been opened for working towards a self-financing climate compact. None of the other types of financial resources available to developing countries have a potential of this scale,' he added.

8 June 2007

⇨ The above information is reprinted with kind permission from OneWorld.net. Please visit www.oneworld.net for more information.

UK legislation: Climate Change Bill

From the UK Department for Environment, Food and Rural Affairs

Background

The debate on climate change has shifted from whether we need to act to how much we need to do by when, and the economic implications of doing so. The time is therefore right for the introduction of a strong legal framework in the UK for tackling climate change. The draft Climate Change Bill is the first of its kind in any country.

The proposed Bill provides a clear, credible, long-term framework for the UK to achieve its goals of reducing carbon dioxide emissions and will ensure that steps are taken towards adapting to the impacts of climate change.

Targets

⇨ This Bill puts into statute the UK's targets to reduce carbon dioxide emissions through domestic and international action by 60% by 2050 and 26-32% by 2020, against a 1990 baseline.

⇨ Five-year carbon budgets, which will require the Government to set, in secondary legislation, binding limits on carbon dioxide emissions during five-year budget periods, beginning with the period 2008-12. Three successive carbon budgets (representing 15 years) will always be in legislation.

⇨ Emission reductions purchased overseas may be counted towards the UK's targets, consistent with the UK's international obligations. This ensures emission reductions can be achieved in the most cost-effective way, recognising the potential for investing in low carbon technologies abroad as well as action within the UK to reduce the UK's overall carbon footprint.

Committee on Climate Change

⇨ A Committee on Climate Change will be set up as an independent statutory body to advise the Government on the pathway to the 2050 target and to advise specifically on: the level of carbon budgets; reduction effort needed by sectors of the economy covered by trading schemes, and other sectors; and on the optimum balance between domestic action and international trading in carbon allowances.

⇨ It will take into account a range of factors including environmental, technological, economic, fiscal, social and international factors, as well as energy policy, when giving its advice.

13 March 2007

⇨ The above information is reprinted with kind permission from the Department for Environment, Food and Rural Affairs. Please visit www.defra.gov.uk for more information.

Climate change is like 'World War Three'

The battle to deal with climate change needs to be fought like 'World War Three', the head of the Environment Agency has warned.

The agency's chief executive Lady Young said current measures to adapt to a changing climate were 'too little, too slowly', and a huge effort was needed to address the crisis.

Hilary Benn, Environment Secretary, warned the agency's annual conference in London that global warming was a challenge to security, migration, politics and economics as well as the environment.

Lady Young told the conference that Britain would face more droughts, flooding, coastal erosion and loss of biodiversity as the climate altered.

She said measures such as improving the resilience of existing homes to flooding, not building on floodplains and improving water use efficiency were needed.

Rising sea levels and coastal erosion threatened £130 billion worth of property around the coast, with the elderly and poor communities most vulnerable, and seaside settlements must have help adapting, she said.

'This is World War Three – this is the biggest challenge to face the

By Charles Clover, Environment Editor

globe for many, many years. We need the sorts of concerted, fast, integrated and above all huge efforts that went into many actions in times of war.

'We're dealing with this as if it is peacetime, but the time for peace on climate change is gone – we need to be seeing this as a crisis and emergency,' she said.

The Environment Agency's chief executive Lady Young said current measures to adapt to a changing climate were 'too little, too slowly'

She also said much needed to be done to reduce greenhouse gas emissions – or adaptation measures would need to massively increase.

But she criticised the proposed Severn Barrage project – which could generate nearly 5 per cent of the UK's electricity through renewables at the cost of the internationally-important

wildlife sites in the estuary – as looking for paper to write on and 'reaching for the Mona Lisa'.

Robert Watson, chief scientist of the Department for the Environment, Food and Rural Affairs, was asked later whether Britain needed to spend 42 per cent of its budget on climate change as the United States did on the war in World War Two. He said tackling climate change required will but was possible at relatively little cost.

Mr Benn told the conference that there was no more important task for the world than dealing with climate change, for the benefit of future generations.

As the century progressed, people would be fighting not just for ideology but for water, and increasing numbers would be refugees of environmental catastrophes, he warned.

'This is not just an environmental challenge. It's also a security challenge, a migration challenge, a political challenge and an economic challenge as well,' he said.

At home, lessons had to be learned from the summer's flooding such as dealing with surface water and confusion in who controls drainage more effectively.

And while climate change mitigation measures such as a post-Kyoto deal and renewable energy were crucial, the Environment Secretary also called on individuals to take 'small steps' such as changing their light bulbs and walking more, which would add up to a 'powerful movement' for change.

Energy minister Malcolm Wicks insisted the Government was 'fully committed' to back EU targets of 20 per cent energy from renewables by 2020, but said British targets were still being negotiated.

And he echoed Mr Benn's call for individuals to take personal action to adopt measures such as improving energy efficiency.

5 November 2007

© Telegraph Group Ltd, London 2007

Thousands of African climate refugees riot at Italian and Spanish detention centres.

Tensions increase between Israel and Syria over the severe water shortage.

⑧News

Australia's food shortage worsens after a 20-year long drought.

US government lays claim over Antarctic oil resources, and announces force will be used if challenged.

China's air pollution crisis continues. Respiratory diseases overwhelm hospitals.

Tuvalu and other Pacific islands underwater. Millions now homeless.

Right to be suspicious

Climate change cannot be tackled if existing injustices in global politics are overlooked

By William Gumede

Post-G8 report cards are for the most part judging that the emphasis in Germany last week was on climate change, with the fight against poverty in Africa and the developing world taking a back seat. In truth, however, the two are so closely intertwined that they cannot be considered separately. Just as skewed global trade and political systems stack the deck against developing countries struggling to escape the poverty trap, it also limits their scope for effective action on climate change.

Progressive efforts to tackle climate change in Africa and the developing world are almost invariably hamstrung by global political, trade and finance rules and realities. Attempts to crack down on energy leakage are too often stymied simply because the mostly international corporations affected can threaten to pack up and move. Poor countries are desperately dependent on investments and jobs from these western companies.

Many developing countries have high levels of carbon emissions because they use so-called dirty fuel such as coal to generate the bulk of their energy. These countries worry about the cost of rapidly turning to sustainable energy, when they have massive social obligations to their poor citizens. More than 25% of households in South Africa, for example, do not have access to affordable energy, let alone clean energy.

The conversion from dirty to clean fuel is expensive. And here there is a telling echo of struggles for antiviral drugs in Africa: countries pursuing new technology to produce cleaner energy affordably often face battles with western companies and governments over intellectual property rights issues.

The people of Africa and the developing world understandably worry that they will find themselves left bearing the brunt of climate change, just as they have regarding health issues. The latest reports from the UN's Intergovernmental Panel on Climate Change have identified Africa as the continent likely to be hardest hit by climate change, thanks to plummeting food production and water shortages. And yet the industrialised countries are disproportionately responsible for global warming. The big developing countries – China, India and Brazil – are not blameless, but the western track record is hardly an example to follow.

After the G8 meeting, many welcomed the news that the United States had agreed that a future deal on the environment would be cobbled together under the auspices of the United Nations. However, the UN is viewed by many in Africa with distrust, especially following its apparent manipulation by the US and 'coalition of the willing' in the lead-up to the invasion of Iraq. There is little confidence that a fair deal will be agreed. At the UN-sponsored Africa climate change event in Kenya last year, Africans were watching powerlessly from the margins, as they were excluded from discussions that concerned them most.

If the G8 is serious about climate change in Africa and the developing world, one proposal is to refocus the World Bank to help poor nations overcome the cost of shifting to clean energy. Only the G8 nations have the power to achieve that.

It is no wonder that the large developing countries are suspicious of western attempts so far to persuade them to opt for a greener, and more costly, option to catch up with the west. Indeed, some developing countries perceive the clamour over climate change as an attempt by the west to dominate the world's depleting energy sources. Others, such as China, India and Brazil, suspect an ulterior motive on the part of a western world anxious about their high growth rates. These positions may be wrong, but they are certainly understandable.

Global warming has a disproportionate impact on poor countries, but it is, almost by definition, a pressing issue everywhere and for everyone. It cannot, however, be tackled in isolation, divorced from the other problems facing Africa and the developing world. Rich nations would be foolish to imagine that the fight against poverty can be postponed in favour of a focus on climate change. The solution to both demands an equitable partnership in decision-making and restoration of trust between the west and the developing world, and that must begin with genuine efforts to change the inequitable global trade, political and financial systems.

12 June 2007

© Guardian Newspapers Ltd, 2007

Reducing climate change

Respondents were asked 'Which of the actions on this list, if any, do you think will do the most to help reduce climate change?'

Action	%
Recycling	40%
Developing cleaner engines for cars	34%
Avoiding creating waste in the first place	22%
Making fewer car journeys	17%
Using less electricity	16%
Taking fewer foreign holidays	11%
Using public transport	10%
Walking or cycling	10%
Buying locally-grown food	7%
Using water sparingly	4%
Reusing bottles/containers	4%
People having fewer children	4%
Buying organic produce	1%
None of these	2%
Don't know	3%

Source: Ipsos MORI, 4 July 2007.

Carbon offsetting – frequently asked questions

Information from the Department for Environment, Food and Rural Affairs

What is carbon offsetting?

Our everyday actions consume energy and produce carbon dioxide emissions, for example driving a car, heating a home or flying. Offsetting is a way of compensating for the emissions produced with an equivalent carbon dioxide saving. In this way it lessens the impact of a consumer's actions. However, it is important to note that offsetting does not actually reduce the emissions contributing to climate change which is why it is important that we all reduce and avoid consuming energy.

Our everyday actions consume energy and produce carbon dioxide emissions, for example driving a car, heating a home or flying

---

Carbon offsetting involves calculating your emissions and then purchasing 'credits' from emission reduction projects that have prevented or removed the emission of an equivalent amount of carbon dioxide elsewhere.

⇨ Due to the fact that greenhouse gases have a long life-span and tend to mix evenly in the atmosphere it doesn't matter where gases are emitted in the world: the effect on climate change is the same. It is easier to reduce emissions from some sources than others but sometimes emissions are unavoidable. To make up for unavoidable emissions increases, equivalent emissions reductions can be made elsewhere, meaning that the overall effect is zero.

⇨ Emission reductions can be made by investment in tech-nology projects, e.g. in renewable energy and energy efficiency. For example, a fossil fuel burning generator could be replaced with a wind turbine, or a community could be fitted with solar water heaters and insulation to reduce its energy use and therefore produce lower carbon dioxide emissions.

Will offsetting solve climate change?

Government acknowledges that carbon offsetting is not a cure for climate change but it can help raise awareness and reduce the impact of our actions. The most appropriate action to take is to reduce emissions.

Offsetting is a useful element of what we can all do to address climate change for several reasons:

⇨ Providing the means to calculate emissions attributable to our activities helps raise awareness of our own impact on climate change. Combined with reducing our emissions, offsetting can be used to address this impact.

⇨ When done in a robust and responsible manner, for example through the purchase and cancel-lation of Certified Emission Reductions (CERs), offsetting leads to a reduction in carbon dioxide emissions in the area local to the offsetting project, often in developing countries. On a global scale, offsetting seeks to maintain a balance between emissions creation and reduction.

⇨ Offsetting projects, such as those approved by the United Nations, provide a mechanism for investment in clean technology in the areas which lack it the most. Such investment can lead to the spread of low-carbon dev-elopment across entire regions, further reducing climate change impact.

What actions can I offset?

Individuals and businesses can offset the fuel used to power and heat their homes or offices, and can also offset their transport emissions from road, rail and air travel. It can be for a particular action, for example a car journey, or activities over a period of time, for example a person's annual mileage. Consumers can offset anything that produces emissions of greenhouse gases.

How can I offset my emissions?

There is a growing number of companies selling offsets. Sometimes these offsets are sold with goods and services, such as flights, electricity or a new car. Once the Code of Best Practice is in place, accredited offsets will be given a Government quality mark. Consumers will be able to look for this mark when offsetting their emissions.

Isn't offsetting just about planting trees?

We are focusing on a technology-based approach rather than planting trees because avoided emissions are preferable to carbon sequestration through forestry.

The use of forests to mitigate global warming is limited due to the fact that the net amount of carbon they absorb could be released again in the future via fire, disease or changes in land use. There are also complexities over determining the amount of carbon they absorb. But we recognise that methodologies for afforestation and reforestation are under consideration by the Clean Development Mechanism Executive Board.

There are many good reasons for supporting forestry other than for carbon offsetting such as significant poverty reduction impacts from promoting the sustainable use of forestry in developing countries.

Why can't I support projects in the UK?

The only internationally agreed framework governing emission reduction projects in developed countries is Joint Implementation (JI). JI is one of the flexibility mechanisms provided for under the Kyoto Protocol, which are designed to help developed countries meet their emission reduction targets. At the present time, the Government has not set up a framework to allow JI projects in the UK.

The main reason for this is the potential for double counting of emission reductions: the majority of possible JI projects would take place within sectors already captured under the EU Emission Trading Scheme, creating the problem of emission reductions generated under that scheme being sold on again as offsets under the banner of a JI project.

For example, if a community in the UK were retrofitted with renewable energy sources and energy efficiency measures, the local supplying power station would experience a drop in demand. Credits from the community project would be sold as ERUs (JI credits) while the power station would have a resulting excess of EUAs to sell off as well, thereby creating a double counting scenario which would result in the offset being recorded twice.

Similarly, UK forestry is included in the calculation of our annual national emissions footprint, counting as a balance against direct sources of emissions. Therefore, any existing trees cannot be sold as offsets as they would not be additional, while new planting could improve the picture of the national emissions balance and reduce the desire for emission reduction action to take place.

Is the Government saying that offsetting makes carbon intensive activities sustainable?

We believe that carbon offsetting is not a substitute for reducing emissions at source but is:

⇨ the 'next best' solution for mitigating remaining emissions from essential activities after all practical steps have been taken to reduce them;

⇨ a means of raising awareness of travel climate change impacts and the choice of less carbon intensive alternatives, and;

⇨ consistent with UK policy that the environmental impact of the aviation industry should be reflected in the cost of air travel.

What does the proposed Code of Practice on offsetting include?

The Code will set standards for:

⇨ Robust and verifiable emission reduction credits;

⇨ Accurate calculation of emissions to be offset, using statistics and factors published for this purpose by the Government;

⇨ Clear information for consumers regarding the mechanism and/or projects supported;

⇨ Transparent pricing;

⇨ Timescales for cancelling credits, and

⇨ Where offered by a company with other goods and services, those companies will offer a 'compulsory choice' for the consumer to offset.

The consultation on the Code was published in January 2007. A summary of the consultation responses was published in July 2007. This summary said that we would proceed with establishing the Code and it would include Kyoto-compliant credits. The summary also said that we are considering whether, and if so how, credits from the non-regulated market (VERs) should be included within the Code.

I have been offsetting my flights under existing schemes. Have I been wasting my money?

Not at all – individuals and providers have helped to blaze a trail. But as the market is expanding it is necessary to provide some clarity. The Code of Best Practice is designed to enable consumers to offset with confidence.

What is the Clean Development Mechanism or CDM?

The CDM is a procedure under the Kyoto Protocol through which developed countries may finance projects in developing countries to reduce emissions of greenhouse gases,

and receive credits for doing so which they may apply towards meeting mandatory limits on their own emissions. Further information can be found on the Kyoto mechanisms pages of the Defra website, or from the UNFCCC website.

What is meant by the Voluntary and Kyoto compliant sector?

⇨ The Voluntary sector refers to emission reduction projects that are developed outside of the UNFCCC Clean Development Mechanism and which provide emission reductions verified by third-party organisations. These are called voluntary reductions because they are purchased by organisations which wish to make voluntary emission reductions for Corporate Social Responsibility reasons, rather than for the purpose of complying to targets under the EU-ETS for example. The VER emission reductions do not have any tradable value.

⇨ The Kyoto compliant sector refers to emission reduction projects and measures provided for under the Kyoto Protocol. These are collectively known as flexible mechanisms and consist of the Clean Development Mechanism (CDM), Joint Implementation (JI) and international emission trading schemes. Countries with Kyoto targets to reduce their emissions can use the flexible mechanisms to help them achieve those targets.

Why did the Government consultation back Clean Development Mechanism (CDM) projects over voluntary schemes?

Government recognises the UNFCCC and the CDM as the highest internationally agreed standard for emission reductions and is choosing to develop a facility to invest in CDM projects. The Government is aware that the number of CDM projects available is limited at present and that offsets are also being offered on the voluntary market. We recognise that the voluntary offset market has value for raising customer awareness of the climate change impact of their activities. We would encourage those in the voluntary sector to adopt measures such that the assessment of their emission reductions mirrors the CDM process to ensure the integrity of the voluntary offsets being offered.

The voluntary offsetting market has proposed a number of standards. Why did the Government's proposals create something new?

A number of different standards are being set up by the voluntary market for offsetting. This can be confusing for consumers. The Government needs to show leadership in what is a growing and diverse marketplace. Due to the cost and time implications involved with regulating the entire voluntary offsetting sector, the consultation identified Kyoto-compliant credits as being best practice rather than creating new mechanisms.

What is the difference between VERs and CERs?

CERs (Certified Emission Reductions) are emission reductions of unit equal to one metric tonne of carbon dioxide equivalent which may be used by Annex I countries towards meeting their binding emission reduction and limitation commitments under the Kyoto Protocol. CERS must come from projects that have been approved by the CDM executive board (a 10-member board which supervises the CDM). As such the integrity of CERs is assured by the UNFCCC process.

VERs (Verified Emission Reductions) are generated by small-scale projects which are assessed and verified by third-party organisations rather than through the UNFCCC. Their purpose is to offset emissions where compliance with binding targets is the not the primary motive. Projects can be relatively cheap and based in the UK, unlike CDM projects which must be hosted in developing countries (Annex II, UNFCCC).

Why do different company websites give different emissions calculations?

Emissions figures can vary due to the different sources of data used by an offsetting provider, and different methodologies employed to calculate emissions. Some companies may calculate actual figures and others may use averages. In the case of aviation, discrepancies may arise depending on the choice of multiplication factor for converting between the number of miles flown and the resulting carbon dioxide emissions, and whether or not the added effect of emissions at altitude are taken into consideration.

The Code of Practice for offsetting providers will set a standard for the accurate calculation of emissions by households and transport. A database of Government-recognised statistics and factors will be drawn up and published annually.

Why do different companies give different prices to offset emissions from the same flight?

In addition to the conversion factor above, there are variations in the price of offset credits. Contributions to this cost come from the different types of project supported, i.e. in type and in scale, plus an added transaction or service fee paid to the offset provider. For several organisations, this leads to a minimum transaction fee for shorter flights.

Government seems to be pressing very strongly for offsetting in the aviation sector, why is this?

Offsetting is not a substitute for the Government's wider policy on aviation. We believe that the best way of ensuring that aviation contributes towards the goal of climate change stabilisation will be through a well-designed, open international emissions trading regime. We are however seeking to take advantage of the infrastructure in place for the EU ETS and show EU leadership by pressing for the inclusion of aviation as soon as possible. Offsetting is a complementary interim measure for tackling the climate change impacts from aviation, which could be promoted to the wider travelling public to be taken up on a voluntary basis.

Does the Government believe that carbon offsetting for road transport is the way forward in curbing carbon dioxide emission growth from the transport sector?

The Government believes that carbon offsetting can help raise awareness of the impact of road transport on climate change and can be used in addition to other actions to actively tackle carbon emissions from travel. It is important though that offsetting is not used as an alternative to other actions that individuals can take to reduce their emissions. We would look to offsetting companies/ providers to help communicate to their customers the importance of actually reducing their own carbon emissions before turning to offsetting. Offsetting should therefore form part of a systematic approach to making more sustainable travel choices:

⇨ reducing the need to travel (e.g. through the use of video-conferencing). If this is not possible then;

⇨ using the most sustainable form of transport (i.e. trains and/or buses) when travelling is absolutely necessary;

⇨ encouraging consumers to use the personal choices available to them in the best possible way, including purchasing a low carbon vehicle, car sharing and eco-driving;

⇨ mitigating remaining carbon dioxide emissions through carbon offsetting.

⇨ The above information is reprinted with kind permission from the Department for Food, Environment and Rural Affairs. Please visit www.defra.gov.uk for more information.

© Crown copyright

Rockin' all over the world . . .

. . . but just watch your carbon footprint. Live Earth beams 150 acts to an audience of 2bn – and burns more carbon than 3,000 Britons do in a year. Critics and fans are split, but everybody wants to save the world

By David Smith

It was a once-in-a-lifetime event. Yes, another one. Its organisers said the biggest emergency in world history demanded the biggest performance in world history. Sceptics said it would generate more heat than light.

Live Earth rocked around the globe yesterday, with 150 acts performing on all seven continents in a bid to create a 'tipping point' in public consciousness and action against climate change. Al Gore, former US Vice-President turned environmental crusader, claimed one in three people on the planet was watching. The man who lost the US Presidency enjoyed a few minutes as de facto president of the world as he broadcast to 2 billion people from Washington: 'Not many years from now, our children and grandchildren will ask one of two questions, looking back at us in 2007.

'Either they will ask: "What were they thinking, didn't they hear the scientists, see the evidence, didn't they care, or were they too busy?" Or they will ask the second question, which I prefer. I want them to ask of us: "How did they get their act together to successfully solve the climate crisis"?'

Gore and his team were determined to muster enthusiasm for their one-world, 24-hour epic despite yawns of concert fatigue after Live Aid, Live8 and, just six days before, the Concert for Diana. Bob Geldof had complained: 'We are all f*****g conscious of global warming.' But his Live8 co-organiser, Richard Curtis, was at the London concert and, unlike those earlier spectaculars, Live Earth was reaching into China, now the biggest producer of carbon emissions .

Live Earth had also been accused of hypocrisy by bands including Arctic Monkeys, The Who and Muse, who dubbed it 'private jets for climate change'. Its total carbon footprint, including the artists' and spectators' travel and energy consumption, was likely to have been at least 31,500 tonnes, said John Buckley of Carbonfootprint.com – more than 3,000 times the average Briton's annual footprint. One viewer of BBC2's *Newsnight* complained online: 'Would you hold a hog roast to promote vegetarianism?'

Clearly aware of the backlash, the comedian Eddie Izzard, on stage at Wembley, admitted: 'A lot of stuff we tell you we have to do as well. We're probably more guilty than anyone with all this flying around and stuff.'

Ticket sales proved patchy at some of the host cities but Wembley shifted all 65,000 at £55 a go.

'Hello Wembley!' screamed radio DJ Chris Moyles, the show's opening host, against a backdrop of a world map made from the painted tops of oil barrels. 'It's a nice, easy, simple question, can you help save the Earth?' A murmur from the crowd. Moyles said: 'We might be screwed if that's the response.' There was always going to be tension between the light entertainment messenger and the deadly serious message of environmental chaos and catastrophe. Moyles acknowledged as much after asking people to send text messages in exchange for advice on reducing their carbon footprint. 'Serious stuff over!' he said. 'Shall we get back to the show?'

The serious stuff was far from over, and at times it could be earnest. There was an emotive combination of music with footage of elephants, polar bears and coral reefs. Boris Becker tried to interest the audience in German recycling methods. Sets were punctuated by short films urging people to use energy-saving light bulbs.

In a symbolic gesture, Wembley switched off its non-essential lights for a minute's darkness. Then the main attraction, Madonna, who had written a song for the occasion, brought the show to a close with characteristic brio.

The other major concerts were in Sydney, Tokyo, Johannesburg, Shanghai, Hamburg, New York and Rio.

Can rock and roll save the world? was a question much asked yesterday. As Live Earth's organisers are the first to admit, it will take longer than 24 hours to find out.

8 July 2007

© *Guardian Newspapers Ltd, 2007*

Can algae save the world?

Information from the Science Museum

Scientists are trying to turn tiny aquatic plants – algae – into a new sort of fuel... But why? The fuels we currently use are made from coal, oil and gas, which are releasing greenhouse gases like carbon dioxide (CO_2) into the atmosphere. The vast majority of climate scientists are convinced that people are causing the climate to change as a result of this.

Climate change will affect us all – and it's already started. It isn't just about the planet getting warmer. As well as causing extreme weather – such as floods, droughts and storms – climate change will force people out of their homes, increase disease and have massive economic effects.

We can't stop climate change overnight. In the UK and around the world, people will need to make changes to adapt to the new climate, like strengthening flood defences and breeding drought-resistant crops.

To crack climate change, we'll need politicians and industry to act. But every single person can do something to help slow climate change and become part of the solution, not part of the problem.

Changing our behaviour is one way of helping to tackle climate change, but science has a part to play too. Most technologies aim to cut greenhouse gas emissions – but a few scientists also have visions of keeping the Earth cool by blocking out sunlight.

So which technologies will work best? There are so many options, it's hard to know which technologies to take seriously. But for every new technology, you can make sense of the arguments using three questions...

⇨ Can the technology really slow down climate change?
⇨ Does the technology work in practice?
⇨ Will the technology do more harm than good?

To show you how it's done, we've applied our three questions to current biofuels, and then to a possible future biofuel.

> ### Changing our behaviour is one way of helping to tackle climate change, but science has a part to play too

Biofuels are one of the technologies that could potentially help us cut our emissions of greenhouse gases like carbon dioxide (CO_2). They're controversial. But you can use the three questions to highlight the most important issues and decide where you stand.

Can biofuels really slow down climate change?

Burning fossil fuels releases carbon that's been locked away for millions of years. This is rapidly adding extra CO_2 to our atmosphere today. Burning biofuels also releases CO_2. But this is exactly the same amount of CO_2 the plant absorbed from the atmosphere while it was growing. Perfect? Not quite...

Biofuels have to be cultivated, processed and transported – and using fossil fuels to do all that adds extra CO_2 to the atmosphere. We can call this biofuels' 'carbon footprint'.

Do biofuels work in practice?

Some power stations already use biomass as a fuel, either on its own or with coal.

Modified cars can run on pure bio-diesel or bio-ethanol. Even ordinary cars can run on petrol mixed with a dash of bio-ethanol. The UK government wants 5% of transport fuel sold at garages to be biofuel by 2010, mainly blended with standard fuel. In fact biofuels could be at a forecourt near you already.

But replacing all our energy sources with biofuels would mean growing a huge amount of biomass – and in the UK there isn't enough land.

Will biofuels do more harm than good?

The more land that is used for biofuels worldwide, the less there is for food crops. This has already led to rising food prices in Italy and Mexico. At worst, competition for land could lead to dwindling food supplies and starvation.

In some countries farmers are chopping down rainforests to plant biofuel crops. This threatens wildlife, and releases greenhouse gases by removing the trees and peatlands that lock away CO_2.

On the other hand growing biofuel crops could help each country produce its own energy, and it could also boost some farmers' incomes.

Like all new technologies biofuels have their pros and cons. Scientists are working to improve them for the future – by developing new production processes and investigating alternative fuel crops. One surprise contender is... algae.

Algae are water plants – some species are so tiny they can only be seen under a microscope. Algae absorb CO_2 and store the carbon as oil. By growing algae in large numbers, scientists can create enough oil to make bio-diesel.

Could algae save the world? How do algae biofuels stand up to our three climate-change-busting questions?

Can algae biofuels really slow down climate change?

Algae absorb CO_2 from the atmosphere as they grow – just like other plants – and release the same amount when the algae biofuel is burned. But algae biofuels still have a carbon footprint. Growing and processing algae, then transporting the biofuel, requires energy. If this comes from fossil fuels, it all creates extra CO_2.

One big attraction of algae is they could potentially replace more fossil fuels than today's biofuel crops. They don't need as much space to grow, so we could make more biofuel with the land we've got.

Do algae biofuels work in practice?

The problem is, although small algae factories work well, scientists aren't convinced they'll ever be able to scale them up to full size. Some people argue we shouldn't focus our attention on unproven ideas, when we could fight climate change using technologies we already have.

What's more, to be a realistic alternative to fossil fuels, algae biofuels will need to be affordable. The balance might change as oil becomes scarcer, or if carbon emissions are taxed, but recent estimates suggest algae biofuels will struggle to compete on price.

Will algae biofuels do more harm than good?

Algae need less space than other biofuel crops – and they don't need fertile farming land either. Algae could be grown anywhere in ponds or factories. This might mean less habitat destruction, and less competition between food and fuel.

It's an exciting vision of the future, but there could still be drawbacks. The main fuel from algae would be bio-diesel – tests on today's bio-diesel show an increase in some harmful air pollutants.

And as algae biofuels are still under development, there could be other complications we just don't know about yet.

As with most new technologies, the arguments surrounding biofuels are complex. We've used three questions to help you consider the issues, weigh up the pros and cons and decide what you think. Do you think algae can save the world?

You can use these questions to help you weigh up any other technology that claims to slow down climate change. And there are lots out there.
⇨ Will it really slow down climate change?
⇨ Is it really possible?
⇨ Will it have bad effects?

Technology alone can't solve climate change, but it can help. Scientists think, rather than pinning our hopes on one idea, we'll need to develop a combination of technologies.

⇨ The above information is reprinted with kind permission from the Science Museum. Please visit www.sciencemuseum.org.uk for more.

Oceans offer climate cure

Two leading environmental scientists have claimed that the oceans could hold the answer to slowing down climate change. By Kate Martin

In a letter to the journal *Nature*, James Lovelock, author of the Gaia Theory, and Chris Rapley, director of the Science Museum, have proposed using vertical pipes in the ocean to help marine plantlife absorb more carbon dioxide.

The pipes, which they suggest could be 100 to 200 metres long and 10 metres in diameter, would use wave movement to pump colder, nutrient-rich water to the ocean surface to fertilise algae in the surface waters.

These plants would consume carbon dioxide through photosynthesis and could also produce chemicals that help to cool the climate through cloud formation.

In their letter to *Nature*, the scientists said: 'Measurements of the climate system show that the Earth is fast becoming a hotter planet than anything yet experienced by humans.

'Processes that would normally regulate climate are being driven to amplify warming.

'Such feedbacks, as well as the

e d i e
environmental data
interactive exchange

inertia of the Earth system – and that of our response – make it doubtful that any of the well-intentioned technical or social schemes for carbon dieting will restore the status quo.

'What is needed is a fundamental cure.'

They added: 'The removal of 500 gigatonnes of carbon dioxide from the air by human endeavour is beyond our current technological capability.

'If we can't heal the planet directly,

we may be able to help the planet heal itself.'

The technology is already being developed by American firm Atmocean, using pipes 200 to 300 metres long and three metres wide.

Company bosses say the pipes could sequester an additional two billion tons of carbon per year on the ocean floor and stabilise carbon dioxide levels in the atmosphere below 550 parts per million.

They believe the process will be helped by zooplankton species which consume algae and excrete pellets containing carbon dioxide that rapidly sink to the seabed.

Their research suggests that the pipes, which bring cooler water to the ocean's surface, could also reduce the intensity of hurricanes.
28 September 2007

⇨ The above information is reprinted with kind permission from edie.net. Please visit www.edie.net for more.

Carbon sequestration

Information from the Forestry Commission

Forest ecosystems make an important contribution to the global carbon budget. This is because of their potential to sequester carbon in wood and soil but also because of their potential to release it if forests are cleared. Many countries and organisations, including the UK Government and the Forestry Commission, are cautious about promoting carbon sequestration as a means of reducing atmospheric carbon dioxide. The size of the potential gains is uncertain and the accounting procedures complicated. Moreover, there is a limit to the amount of carbon that woodland can sequester, and there is a risk that the sequestered could be released – through, for example, felling, forest fires or outbreaks of pests and diseases.

Forests and woodlands in the UK contain around 150 million tonnes of carbon, and every year they remove about 4 million tonnes of carbon from the atmosphere. These values need to be compared with total UK emissions of around 150 million tonnes of carbon (as carbon dioxide) every year – mainly due to the combustion of fossil fuels. So the forest carbon sink is offsetting less than 3% of annual carbon dioxide emissions and the accumulated carbon stock in forests represents only about one year's worth of emissions at current rates. The rate of carbon sequestration (4 million tonnes per year) is relatively high because most of the UK's forests are young and still growing. As our forests grow older, the rate of carbon dioxide removal will fall. However, carbon sequestered in UK forests is still an add-on to the many other benefits that can arise from forestry as long as the forestry authorities maintain a clear commitment to sustainable forest management.

How can trees and forests act as a carbon sink?

The term 'sink' is used to mean any process, activity or mechanism that removes a greenhouse gas from the atmosphere. Forests and other green vegetation exchange large amounts of greenhouse gases with the atmosphere. Plants capture carbon dioxide from the atmosphere through photosynthesis, releasing oxygen and part of the carbon dioxide through respiration, and retaining a reservoir of carbon in organic matter. If stocks of carbon are increased by afforestation or reforestation, or carbon stocks in croplands or forest stands are increased through changes in management practices, then additional carbon dioxide is removed from the atmosphere. For example, if an area of arable or pasture land is converted to forest, additional carbon dioxide will be removed from the atmosphere and stored in the tree biomass. The carbon stock on that land increases, creating a carbon sink. However, the newly created forest is a carbon sink only while the carbon stock continues to increase. Eventually an upper limit is reached where losses through respiration, death and disturbances such as fire, storms, pests or diseases or due to harvesting and other forestry operations equal the carbon gain from photosynthesis. Harvested wood is converted into wood products and this stock of carbon will also increase (act as a sink) until the decay and destruction of old products matches the addition of new products. Thus a forest and the products derived from it have a finite capacity to remove carbon dioxide from the atmosphere, and do not act as a perpetual carbon sink. By substituting for fossil fuels, however, land used for biomass and bioenergy production can potentially continue to provide emissions reductions indefinitely.

If a forest area is harvested and not replanted, or is permanently lost due to natural events like fire or disease, then the carbon reservoir that has been created is lost. In contrast, the

benefits provided by bioenergy substituting for fossil fuels are irreversible, even if the bioenergy scheme only operates for a fixed period. Frequently a distinction is made concerning the so-called 'permanence' of measures based either on carbon sinks or on replacement of fossil fuel with bioenergy.

Does tree harvesting cancel out the carbon sink?

Forests and woodlands managed for commercial wood production through periodic harvesting generally have lower carbon stocks than stands that are not harvested, but this harvesting should not be confused with deforestation. Deforestation implies a change in land cover from forest to non-forest land, whereas sustainable wood production involves cyclical harvesting and growing. A newly created forest managed for wood production can act as a carbon sink just as surely as a newly created forest reserve, although there may be differences in the level of the ultimate carbon stock and the time horizon over which it is attained.

⇨ The above information is reprinted with kind permission from the Forestry Commission. Please visit www.forestry.gov.uk for more information.

Fuelling the future

Why DFID should ditch dirty development and lead the way to a low carbon future

Ditch Dirty Development is the latest of People & Planet's climate change campaigns. It aims to end the contradiction between government statements and targets to tackle climate change by cutting greenhouse gas emissions, and the continued use of development aid to support fossil fuel extraction projects.

Development aid, earmarked for poverty alleviation, is being put towards some of the most polluting industries in the world in order to feed the energy addiction of the wealthy industrial world. This in turn contributes to climate change, one of the biggest threats to development the world has seen. People & Planet believes that this is a misuse of development money which must stop.

The solutions to climate change exist, primarily in curbing the consumption of fossil fuels in the developed world, and in technologies which harness unlimited and unpolluting renewable energy. In addition, there is a need for strategies to address the growing energy needs of emerging and developing economies without concurrent increases in carbon emissions.

There is a need for boldness of political leadership, and for new thinking about development. Lifting people out of poverty does not have to rely on a business-as-usual approach based on the wasteful use of climate-changing fossil fuels and resource-intensive economic growth. Instead, governments like the UK must support new, low carbon forms of development, which end poverty without contributing to climate change.

DFID inconsistent on energy and climate change

The Department for International Development (DFID) 2006 White Paper states that 'climate change poses the most serious long-term threat to development and the Millennium Development Goals'. This paper documents how, despite this awareness, DFID:

⇨ contributes to climate change through multilateral funding for oil and gas extraction projects. Oil and gas drilling projects funded by international development banks and export credit agencies since 1992 will be responsible for 8.5 billion tons of carbon equivalent over the course of their lifetimes.

⇨ opposed the phase-out of World Bank investment in oil extraction projects.

⇨ has no targets to increase support for renewable energy.

⇨ has no targets to increase access to energy in developing countries.

⇨ does not monitor the impact of its funding on the climate.

In short, DFID's approach to climate change and energy is inconsistent and contradictory.

DFID must promote low carbon development

People & Planet is calling on DFID to produce an energy and climate change strategy, covering both bilateral and multilateral energy funding, which will:

1. Ensure DFID's activities actively contribute to mitigating climate change through global emissions reductions. To do this, DFID must:
⇨ phase out all support for fossil fuel extractive projects.
⇨ massively increase support for new renewable energy sources.

2. Increase access to energy in the developing world, by promoting and championing decentralised and low carbon forms of energy as appropriate.

The strategy must have specific, timebound targets and built-in monitoring systems. DFID should report to Parliament on progress towards these targets on a regular basis.

An energy and climate change strategy would enable DFID to take a leading role in promoting low carbon development. Energy policy is critical both to mitigating climate change and to alleviating poverty, and should be central to DFID's work.

October 2006

⇨ The above information is reprinted with kind permission from People & Planet. Please visit http://peopleandplanet.org for more.

© People & Planet

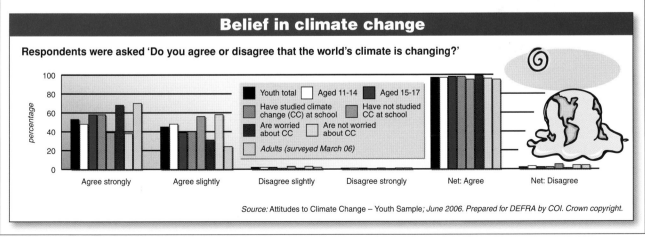

Belief in climate change

Respondents were asked 'Do you agree or disagree that the world's climate is changing?'

Legend: Youth total | Aged 11-14 | Aged 15-17 | Have studied climate change (CC) at school | Have not studied CC at school | Are worried about CC | Are not worried about CC | *Adults (surveyed March 06)*

Categories: Agree strongly | Agree slightly | Disagree slightly | Disagree strongly | Net: Agree | Net: Disagree

Source: Attitudes to Climate Change – Youth Sample; *June 2006. Prepared for DEFRA by COI. Crown copyright.*

Sustainable fossil fuels

The unusual suspect in the quest for clean and enduring energy.
Synopsis of a book by Mark Jaccard

More and more people believe we must quickly wean ourselves from fossil fuels to save the planet from environmental catastrophe, incessant oil conflicts and economic collapse.

This view is epitomised by the claim in one of many recent anti-fossil fuel books that 'Civilization as we know it will come to an end sometime in this century unless we can find a way to live without fossil fuels.'

This view is misguided. This book explains why.

Those who argue that the end of fossil fuels is nigh usually start with evidence that we consume conventional oil faster than we find it, and then link this to the latest energy price spike and geopolitical conflict. What they overlook is that a peak in the production of 'conventional oil' is unlikely to be of great significance given the potential for substitution among the planet's enormous total resources of conventional and unconventional oil, conventional and unconventional natural gas, as well as coal.

A refined petroleum product like gasoline can be produced from any of these other fossil fuels, and indeed it is today from unconventional oil in the form of oil sands (Canada), natural gas (Qatar) and coal (South Africa).

The planet has perhaps 800 years of coal at today's use rate and an even longer horizon for natural gas if we exploit untapped resources like deep geopressurised gas and gas hydrates. While this substitution potential does not mean that energy supply markets will always operate smoothly – prices can oscillate, sometimes dramatically, from one year or decade to the next – it suggests that we should not misinterpret periods of high prices as indicating the imminent demise of our still plentiful fossil fuel resources.

When it comes to fossil fuels, those worried about resource exhaustion find common cause with those worried about environmental impacts. But we can use fossil fuels with lower impacts and less risk. Fossil fuels are a high quality form of stored solar energy – the result of millions of years of photosynthesis that grew plants and the animals that fed upon them, both of whose decomposing remains were trapped in sediments and eventually transformed through subterranean pressures into natural gas, oil and coal.

When humans are ignorant or uncaring about the impacts of using this source of energy, they can create great harm to themselves and the environment. Open pit coalmines destroy mountains and valleys.

Oil spills soil coastlines and harm wildlife. Uncontrolled burning pollutes the air in homes and cities, acidifies lakes and forests, and risks major climate disruption.

This litany of impacts and risks presents a black image for fossil fuels. However, the history of fossil fuel use is also one of humans detecting and then successfully addressing its environmental challenges. Industrialised countries are the most dependent on fossil fuels, and yet in these countries indoor air quality is excellent compared to all of human history since the discovery of fire (with a huge benefit for life expectancy), urban air quality is better in many cities than it was 100 years ago, and acid emissions have fallen in some regions by over 50% in the past 30 years.

The latest challenge is CO_2 emissions from fossil fuel combustion, the most significant of the human-produced greenhouse gases that threaten to raise global temperatures and disrupt weather patterns and ecosystems. But in the decade or so that researchers have grappled seriously with this challenge many promising solutions have appeared. Fossil fuels can be converted to clean forms of energy – electricity, hydrogen and cleaner-burning synthetic fuels like methanol and dimethyl ether – through gasification processes that enable the capture of carbon and its safe storage, most likely deep in the earth's sedimentary formations.

There are costs. Estimates from independent researchers suggest that zero-emission fossil fuel production of electricity would increase final electricity prices by 25-50% were this technology to become universally applied. Researchers also suggest that the cost of vehicle use would increase by about the same percentage as we shifted from gasoline and diesel to primarily hydrogen, electricity and some biofuels for personal mobility. This increase, which is less than recent price jumps of electricity,

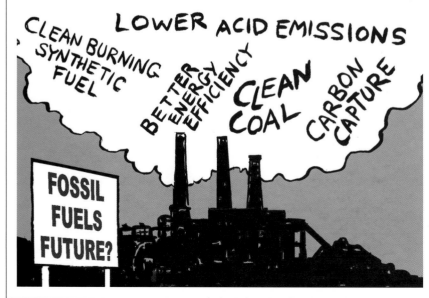

gasoline, heating oil and natural gas in many jurisdictions, implies that the cost of energy would climb over the next century from its current level of 6% to about 8% of a typical family's budget in an industrialised country – remaining much lower than what it was 100 years ago and than what it is today for a poor family in a developing country. Thus, to shift our use of fossil fuels to these zero-emission processes over the course of this century would result in real energy price increases of much less than 1% per year during the next three to five decades.

We can use fossil fuels with lower impacts and less risk

Even if we can use fossil fuels cleanly, however, we might prefer to switch to other options sooner in order to ensure that our energy system is not a house of cards that collapses when we deplete our lower cost fossil fuels. But this decision requires careful consideration of the difficulties and costs involved in forcing the switch quickly versus allowing it to occur gradually as the cost of fossil fuels trends upward in the distant future. We need to have a realistic view of the other options – the 'usual suspects' of energy efficiency, nuclear power and renewable energy.

Energy efficiency has great potential according to physicists, engineers and environmentalists. There are, however, significant countervailing factors that will hinder efforts to make dramatic efficiency gains.

First, efficiency gains lower the cost of energy services and therefore incite some greater use of existing technologies, and especially the innovation and commercialisation of related energy-using technologies, a feedback effect that has caused energy consumption in industrialising countries to grow almost as fast as economic activity over the past two centuries. Ambitious increases in energy efficiency require a dramatic rise in the cost of energy in order to

prevent the widespread adoption of the myriad of energy-using innovations commercialised every year. But, as noted, the shift to zero-emission energy supplies, whether from renewables, nuclear or fossil fuels, is unlikely to increase final energy prices by more than 25-50% from current levels.

Second, the energy system itself will consume increasing amounts of energy in the process of converting lower quality and less accessible primary energy sources (unconventional oil, unconventional gas, coal, and renewables) into higher quality and cleaner secondary energy (electricity, hydrogen and synthetic fuels). The net effect is to decrease the overall efficiency of the global energy system.

Third, the more than 50% increase in the world's population over this century will happen mostly in poorer regions of the world, where energy use is minimal. Even a marginal increase in energy use by people in these countries to provide the most basic services has a profound implication for aggregate energy use at the global scale. Thus, while we should pursue energy efficiency, it is likely that the global energy system will still expand three- or fourfold over this century, especially as people in developing countries use their rising incomes to enjoy energy services that most people in wealthier countries take for granted.

Nuclear power is potentially inexhaustible, but it must overcome public fears about radiation leaks from operational accidents, waste storage and even terrorist attacks, as well as superpower concerns about nuclear weapons proliferation. To make significant advances, therefore, nuclear must be substantially cheaper

than its competitors for providing zero-emission energy. Most cost estimates, however, suggest that it will be cost-competitive at best and perhaps more expensive when the full costs of facility decommissioning, waste disposal and insurance liability are accounted for. The use of nuclear will grow in some regions, but for the next 100 years its share of the global energy system is unlikely to expand much beyond its current 3%. In the more distant future, developments in fusion technology or other non-radiating nuclear alternatives may expand its opportunities.

Renewable energy is seemingly inexhaustible and environmentally benign, yet many of its manifestations are characterised by low energy density, variability of output and inconvenient location. This will often require dedicated facilities for energy concentration, storage, and transmission, and these can cause significant environmental and human impacts depending on their character and scale. The dams and reservoirs of large hydropower projects flood valuable valley bottoms and impede migratory fish and animals, windpower farms can conflict with scenic, wildlife and other values, and biomass energy plantations compete for fertile land with agriculture and forestry. As the contribution of renewables grows in scale, the associated energy concentration and storage costs will become more of an issue. Even when helped by strong policies, it takes time for renewables-using innovations to achieve the commercialisation and expanded production that is necessary to lower costs. Starting from the negligible market share of renewables today, and in a growing global energy system, it will be an

BWEA

enormous and likely very expensive endeavour to force the wholesale replacement of fossil fuels with a renewables-dominated system in the course of just one century.

In anticipating the relative contribution of each of these energy options over this century, it is important not to confuse means and ends. The end is not an energy system dominated by renewables or nuclear or fossil fuels. The end is a low impact and low risk energy system that can meet expanded human energy needs indefinitely and do this as inexpensively as possible, without succumbing to cataclysmic forces at some future time. With this sustainable energy system as the goal, it is unjustifiable to rule out fossil fuels in advance of a holistic comparison that considers critical decision factors. These factors include cost, of course, but also the general human desire to minimise the risk of extreme events (like a major nuclear accident), to ensure adequate and reliable energy supplies free from geopolitical turmoil, and to sustain values, institutions and lifestyles.

Even though it will perhaps triple in size over this century, the global energy system should nonetheless reduce its environmental impacts and risks. If the costs are not too great – and they appear not to be – it can become in effect a zero-emission energy system with negligible impacts to land, air and water. And any residual, unavoidable hazards can be ones from which the system could recover within a reasonable time, either from natural processes alone or in concert with human remediation efforts.

This sustainability objective for the global energy system is achievable, and indeed we have several options. But when all of these options are compared without prejudice, fossil fuels – the 'unusual suspect' – are likely to retain a significant role in the global energy system through this century and far beyond, and the transition toward renewables and perhaps eventually nuclear will be gradual. Deliberately diverting from this lowest cost path by prematurely forcing fossil fuels out of the energy supply mix may not mean as much for wealthy countries,

but for the poorer people on this planet this arbitrary requirement would divert critical resources that could otherwise be devoted to essential investments in clean water, health care, disease prevention, education, basic infrastructure, security, improved governance and biodiversity preservation.

Ironically, however, clean energy – whether relying on fossil fuels or some other option – does not ensure a sustainable human presence on earth. Indeed, if the eventual, long-

term costs of developing a clean energy system are as low as some of the evidence suggests, the challenges to sustainability may be even greater as humans use energy to satisfy their basic needs and seemingly inexhaustible desires for materials and living space.

⇨ Information from the Energy and Materials Research Group at Simon Fraser University. Please visit www.emrg.sfu.ca for more information.
© *Simon Fraser University*

Climate change and the need for renewable energy

Information from the BWEA

The amount of carbon dioxide (CO_2) in the atmosphere has been rapidly increasing over the last 100 years. This is due to the burning of fossil fuels like coal and oil, which contain carbon. As the proportion of CO_2 in the atmosphere changes, the way it retains heat also changes. Scientists now believe this is what is causing the average temperature of the earth to increase, leading to climate change.

Many people predict that the rapid rise in damage caused by natural disasters over the last 30 years is linked to climate change, and unless the global community changes the way it uses and generates energy this process may completely disrupt the global economy in years to come, along with countless lives. In response to this threat, the UN agreed the Kyoto Protocol in Japan in 1997. This requires industrialised nations to reduce greenhouse gas emissions (CO_2 and others) by 5% of 1990 levels by 2008-2012. The UK has agreed to meet this target, but this is only the start of what is needed.

In the Government's recent white paper it stated its goal of a 50% reduction in carbon emissions by 2050. This is a massive task. Renewable energy is the name for technologies which can generate electricity or heat people's homes without burning fossils fuels. The energy source is constantly renewed, like the energy in the wind or waves. Getting to this target will mean a huge increase in the amount of renewable energy we use.

The Government has already put in place legislation to begin this process. The Renewables Obligation came into force in April 2002, obliging all electricity suppliers to source 10% of their supply from renewable technologies by 2010. Wind energy is uniquely placed to start this process and is a rapidly expanding business, set to account for 8% of electricity generation by 2010. It is by far the cheapest of the renewable energy technologies, already generating electricity at prices that are competitive with conventional thermal power plants and well below those for coal and nuclear.

⇨ The above information is reprinted with kind permission from the BWEA. Please visit www.bwea.com for further information.
© *BWEA*

Combat climate change with fewer babies

Information from the Optimum Population Trust

A radical form of 'offsetting' carbon dioxide emissions to prevent climate change is proposed today – having fewer children.

Each new UK citizen less means a lifetime carbon dioxide saving of nearly 750 tonnes, a climate impact equivalent to 620 return flights between London and New York,* the Optimum Population Trust says in a new report.

A radical form of 'offsetting' carbon dioxide emissions to prevent climate change is proposed today – having fewer children

---

Based on a 'social cost' of carbon dioxide of $85 a tonne,** the report estimates the climate cost of each new Briton over their lifetime at roughly £30,000. The lifetime emission costs of the extra 10 million people projected for the UK by 2074 would therefore be over £300 billion. ***

A 35-pence condom, which could avert that £30,000 cost from a single use, thus represents a 'spectacular' potential return on investment – around nine million per cent.

The report adds: 'The most effective *personal* climate change strategy is limiting the number of children one has. The most effective *national* and *global* climate change strategy is limiting the size of the population.

'Population limitation should therefore be seen as the most cost-effective *carbon offsetting strategy* available to individuals and nations – a strategy that applies with even more force to developed nations such

as the UK because of their higher consumption levels.'

A *Population-Based Climate Strategy*, the OPT's latest research briefing, is published today (Monday, 7 May 2007). It says human population growth is widely acknowledged as one of the main causes of climate change yet politicians and environmentalists rarely discuss it for fear of causing offence. The result is that a 'de facto taboo' exists, throughout civil society and government.

One consequence is that 'couples making decisions about family size do so in the belief that it is a matter for them and their personal preferences alone: the public debate and awareness that might have encouraged them to think about the implications of their choices for

their fellow citizens, the climate and the wider environment have been missing.'

Other points in the briefing include:

⇨ Providing low-carbon electricity for the 11 million extra UK households forecast for 2050 would mean building seven more Sizewell B nuclear power stations or 10-11,000 wind turbines.

⇨ Global population growth between now and 2050 is equivalent in carbon dioxide emissions terms to the arrival on the planet of nearly two more United States, over two Chinas, 10 Indias or 20 UKs.

⇨ Even if by 2050 the world had managed to achieve a 60 per cent cut in its 1990 emission levels, in line with the Intergovernmental

Tackling climate change

Respondents were asked 'Which, if any, of the following do you think will have the most impact on you personally if climate change were successfully tackled?'

- A cleaner atmosphere — 58%
- Less severe weather — 30%
- Slower spread of disease — 15%
- Lower insurance premiums — 5%
- Greater stability and security for my children — 31%
- Not having to pay for the new infrastructure needed to cope with climate change (eg flood defences) — 11%
- Don't have to suffer the discomfort of high temperatures — 9%
- Lower increase in summer smog and air pollution — 20%
- Less likelihood of water shortages — 17%
- Less likelihood of being flooded — 12%
- Greater variety / quality of wildlife or countryside — 22%
- Other — *
- Don't know — 2%

Respondents were asked 'How strongly do you agree or disagree that ...?'

I would do more to try to stop climate change if other people did more, too

- Strongly disagree 12%
- Undecided/don't know 2%
- Strongly agree 14%
- Tend to disagree 21%
- Neither agree nor disagree 13%
- Tend to agree 39%

The government should take the lead in combating climate change, even if it means using the law to change people's behaviour

- Strongly disagree 7%
- Undecided/don't know 2%
- Tend to disagree 10%
- Strongly agree 27%
- Tend to agree 43%
- Neither agree nor disagree 11%

Source: Ipsos MORI, 4 July 2007. Ipsos MORI interviewed a representative sample of 2,031 adults aged 16+ face-to-face, in home, on 14-20 June 2007.

Panel on Climate Change's recommendations and the UK Government's target, almost all of it would be cancelled out by population growth.

It concludes: 'A population-based [climate] strategy…involves fewer of the taxes, regulations and other limits on personal freedom and mobility now being canvassed in response to climate change…To sum up, it would be easier, quicker, cheaper, freer and greener.'

Valerie Stevens, co-chair of the OPT, said:

'We appreciate that asking people to have fewer children is not going to make us popular in some quarters. Equally, expressing concern about the environmental impacts of mass migration, which currently accounts for the bulk of population growth in the UK and will have a major effect on Britain's carbon emissions, is a quick route to being labelled racist. But these are hugely important issues and the unfortunate fact is that both politicians and the environmental movement are in denial about them. It's high time we started discussing them like adults and confronting the real challenges of climate change.'

She added: 'Government fiscal measures that support child-bearing however many children a couple has, send a signal that increasing numbers are good for the welfare of everyone. In a world needing to diminish its consumption of key resources, especially energy, this is sadly no longer true.'

NOTES

*Based on 1.2 tonnes of carbon di-oxide per return flight (Department for Transport).

**Stern Review, October 2006.

***Fertility levels in the UK have been below replacement level (2.1 children per woman) for around 30 years. Inward migration is currently the main driver of UK population growth, accounting for over 80 per cent of projected increase to 2074. However, even without the effects of immigration, demographic momentum – the result of the large numbers of children produced in earlier age bands reaching child-bearing age – would have prevented any population decline up to the present. The total fertility rate (TFR) peaked in 1964 at 2.95 children per woman, but this was followed by a rapid fall in the number of births per woman in the 1970s. In 2005 the TFR in the UK was 1.78 children; it is expected to level off at 1.74 (Office of National Statistics).

7 May 2007

⇨ The above information is reprinted with kind permission from the Optimum Population Trust. Please visit www.optimumpopulation.org for more information.

© Optimum Population Trust

New eco-towns could help tackle climate change

New small zero carbon 'eco-towns' built on brownfield land could lead the way in cutting carbon emissions and building affordable homes, Housing Minister Yvette Cooper said today

New small zero carbon 'eco-towns' built on brownfield land could lead the way in cutting carbon emissions and building affordable homes, Housing Minister Yvette Cooper said today.

The Government announced it would consider plans for eco-towns put forward by local authorities as part of the New Growth Points scheme. Forty-five councils have already come forward with plans for new homes and jobs to respond to serious housing pressures in their areas, and some authorities are also looking at plans for 'new settlements'. Ministers will now consider these plans within the Growth Points scheme, insisting on proposals for zero- or low carbon developments which make the best use of brownfield land.

New eco-towns, of between 5,000 and 10,000 homes, would have strong public transport links to nearby towns and cities. They would make the best use of brownfield land and could be built on public sector surplus land such as former MoD or NHS sites. Ministers believe these new developments could help drive the environmental technologies needed to ensure all new homes are zero carbon within a decade, as set out in last December's zero carbon timetable.

Yvette Cooper today announced £2m funding to develop plans for the eco-towns. She also announced the appointment of Professor David Lock, Chair of the Town and Country Planning Association, to report to Government on further developing the criteria for eco-towns.

Yvette Cooper said:

'We desperately need more homes – and we desperately need to cut carbon emissions to tackle climate change. New eco-towns could build low carbon design into the fabric of the community, not just into individual houses.

'We have already made substantial progress, with the new timetable for zero carbon development and proposals for places like Northstowe.

'But we need to go further. Now is the time for us to look at new eco-towns, put forward by local councils. They could use public transport and new green designs to deliver low cost and low carbon homes for the future, making good use of brownfield land.'

7 March 2007

⇨ The above information is reprinted with kind permission from the Department for Communities and Local Government. Please visit www.communities.gov.uk for more information.

© Crown copyright

The inconvenient truth about carbon offsetting

In the concluding part of a major investigation, Nick Davies shows how greenhouse gas credits do little or nothing to combat global warming

It is 20 months now since British Airways proudly announced a new scheme to deal with climate change: for the first time, passengers could offset their share of the carbon produced by any flight by paying for the same amount of carbon to be taken out of the atmosphere elsewhere. 'I welcome warmly this move from BA,' said the then environment minister, Elliot Morley.

And how much carbon has BA offset from the estimated 27m tonnes which its planes have fired into the air since that high-profile moment in September 2005? The answer is less than 3,000 tonnes, less than 0.01% of its emissions – substantially less than the carbon dispersed by a single day of its flights between London and New York. The scheme has been, as BA's company secretary, Alan Buchanan, put it to a House of Commons select committee earlier this year, 'disappointing'.

The project has failed, according to one well-placed BA executive, because one part of the company wanted to improve its image by going green while another part wanted to protect its image by saying nothing at all about the impact of air travel on global warming. The result was that the scheme was launched and then banished to a dark corner of BA's website.

That tension – between the demands of the planet and the imperatives of commerce – lies at the heart of the global response to climate change and, in particular, of carbon offsetting. The idea that we might cancel our own greenhouse gases by paying for projects that reduce the gases elsewhere was born in the early years of climate politics. It was adopted by the corporate lobby at the Kyoto summit in 1997 and has grown into a large but deeply troubled adolescent – confused, unpredict-

able, and difficult to trust.

Separately from the 'compliance market' on which nations and corporations trade carbon credits in an attempt to hit their Kyoto targets, there has grown a smaller, voluntary market in which airlines, banks, car makers and energy companies queue up to offset their carbon and to encourage their customers to do the same. A *Guardian* investigation suggests that many of the schemes on offer here are well-meaning but thoroughly unreliable.

That tension – between the demands of the planet and the imperatives of commerce – lies at the heart of the global response to climate change

---

One company, Equiclimate, which is run by Christians and recommended by the government, has sold thousands of tonnes of offset which are now worthless in financial and environmental terms. It bought up some of the special permits which allow European companies to emit specified amounts of carbon. The idea was to sell them to customers who would 'retire' them, thus cutting the amount of carbon which those companies could produce. But the European commission distributed 170m too many of the permits and so the thousands which have been bought by Equiclimate's customers make no difference at all. People may believe they are offsetting the

emissions from
their patio heaters by signing up to the Calor Gas offsetting scheme, but the sad fact is that Calor Gas is relying on 5,000 tonnes of EU permits which it bought from Equiclimate when most of the permits were already worthless. 'We chose them because they were recommended by government,' a Calor Gas executive said.

The British government itself has been caught out over emissions from its presidency of the G8 in 2005. The then environment secretary, Margaret Beckett, said that all carbon emissions from all meetings and travel linked with the one-year presidency would be neutralised. Delegates to the Gleneagles summit in July 2005 were given certificates declaring that all their carbon emissions were being offset. But it hasn't happened. The plan was to spend £150,000 in Kuyasa, a suburb of Cape Town in South Africa, refitting shack-like homes with insulated roofs instead of corrugated iron, and providing solar panels for electricity and long-life bulbs for light. But the project, which would cut carbon emissions as well as helping needy people, has run into bureaucratic, financial and practical problems. The environment department, Defra, says

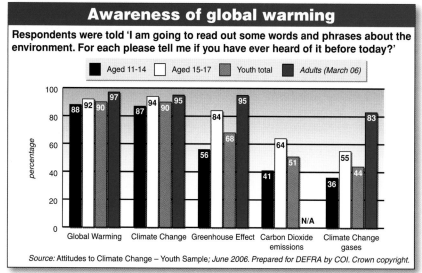

Awareness of global warming

Respondents were told 'I am going to read out some words and phrases about the environment. For each please tell me if you have ever heard of it before today?'

Legend: Aged 11-14 | Aged 15-17 | Youth total | Adults (March 06)

Category	Aged 11-14	Aged 15-17	Youth total	Adults (March 06)
Global Warming	88	92	90	97
Climate Change	87	94	90	95
Greenhouse Effect	56	84	68	95
Carbon Dioxide emissions	41	64	51	N/A
Climate Change gases	36	55	44	83

Source: Attitudes to Climate Change – Youth Sample; *June 2006. Prepared for DEFRA by COI. Crown copyright.*

it is keeping it under review. The project's leader, Shirene Rosenberg, says she is still fighting to keep it alive with no start date on the horizon.

Climate pack withdrawn

Following a phone call from the *Guardian*, the Science Museum has withdrawn its Climate Relief Gift Pack, which included a certificate offsetting 100kg (220.5lb) of carbon and an opportunity to offset a tonne more. The pack promised to 'instantly reduce the amount of CO_2 emitted into the atmosphere and help reduce global pollution'. This was nonsense. The offset, like Equiclimate's, was based on 'retiring' EU carbon permits whose supply easily exceeds demand. It was also overpriced. While the Science Museum was offering them at £30 a tonne, EU permits were for sale at 19p a tonne. The company behind the scheme, Moon Estates of St Austell in Cornwall, also withdrew the product from sale on its website. A company executive admitted they had sold some 3,000 tonnes, at a potential total of £90,000.

Atmosfair, a German offsetting group which is well regarded for its commitment to the environment, undertook to rewrite a section of its website following a phone call from the *Guardian*. Since 2004, it has been offering air travellers offsets which carry the gold standard awarded by a Swiss-based group backed by dozens of environmental NGOs. In an uncertain market, this gold standard is highly desirable. But none of the five projects on which Atmosfair is relying has yet produced

a single verified gold standard reduction in emissions. One project was never intended to reach gold standard; one has been withdrawn; one is stalled. The remaining two – solar-powered kitchens in India and energy from palm oil waste in Thailand – are up and running, but neither has yet completed the gold standard process. Atmosfair's founder, Dietrich Brockhagen, acknowledged that what he was selling was 'forward' credits even though the two projects might fail finally to generate them. 'You have a point, that the customer might not understand this,' he said.

The problem with offsetting is twofold. First, these schemes are unregulated and wide open to fraud. There is nothing but the customer's canniness to stop a company claiming to be running a scheme which does not exist; claiming wildly exaggerated carbon cuts; selling offsets that have already been sold; charging hugely inflated prices. EasyJet, the cut-price airline, backed out of offsetting in April on the grounds that 'there are too many snake-oil salesmen in the business'.

Second, as all the examples above show, even the most well-intentioned schemes suffer from basic weaknesses in the idea of carbon offsetting – an idea which flows not from environmentalists and climate scientists trying to design a way to reverse global warming but from politicians and business executives trying to meet the demands for action while preserving the commercial status quo. It fails on at least three essential points.

First, it requires an accurate measure of the emissions to be offset. That turns out to be riddled with uncertainty. The UN's Inter-governmental Panel on Climate Change found a margin of error of 10% with measuring emissions from making cement or fertiliser; 60% with the oil, gas and coal industries; and 100% with some agricultural processes. Measuring emissions from aircraft is especially fraught with disagreement about what exactly should be measured and aggravated by variations in each flight's height, cargo load and weather conditions. When Tufts University in Maine analysed offsetting websites, it found emissions for flights between Boston and Frankfurt being calculated at anything between 1.43 tonnes and 4.14 tonnes.

Second, it requires an accurate measure of the carbon saved elsewhere. Most of the earliest offset projects involved planting trees, which naturally ingest carbon, a complex and unpredictable process which forbids accurate measurement.

Projects that focus on energy efficiency are even trickier. Carbon Offsets Ltd, another company recommended by the government, is selling offsets from a South African project known as Basa Magogo. This encourages poor households who make coal fires in perforated cans called imbawulas to build the fire in a different way: instead of using paper, then wood with coal on top, they are to build them with most of the coal on the bottom, thus producing more heat and less smoke. But how does anybody check how many have built their fires this way? And how many imbawulas must burn this way for how long before a tonne of carbon is saved? Hugh Somerville, one of the founders of Carbon Offsets Ltd and clearly genuinely keen to offer a decent service, confessed that nobody had asked this question before.

Finally, the very idea of off-setting relies on what is known as additionality – evidence that a carbon reduction would not have occurred in the natural order of commercial life. One of the biggest UK offsetters, Climate Care, which is used by the *Guardian*, distributed 10,000 energy-efficient light bulbs in a South African township; offered the carbon reductions as offsets; and then discovered that an energy company was distributing the same kind of light bulbs free to masses of customers, including their township, so the reduction would have happened anyway.

The result of these fundamental problems is a crisis of legitimacy in the voluntary market, as offsetters lay claim to certainties that are beyond their reach. Dan Welch, a Manchester journalist who investigated offsetters for *Ethical Consumer* magazine, summarised it neatly: 'Offsets are an imaginary commodity created by deducting what you hope happens from what you guess would have happened.'

The early forestation projects, so beloved of rock bands, have been discredited. Apart from their inability to make accurate measurements of carbon saved, companies were offsetting immediate emissions with reductions that would occur only during the 100-year life span of a tree. One of the first, Future Forests, was selling offsets from a forest at Orbost in Scotland. Customers may have thought that they were paying for new trees to be planted, but the company's contract with the forest's owner reveals that all they were paying for was the right to claim ownership of carbon absorbed by trees which were planted anyway. The Advertising Standards Authority in November 2002 ruled that Future Forests had misled customers into thinking that their offset money would be used to grow new trees.

'A waste of time'

Some tree-planting projects in Guatemala, Ecuador and Uganda have been accused of disrupting water supplies; evicting thousands of villagers from their land; seizing grazing rights from farmers; cheating local people of promised income; and running plantations where the soil releases more carbon than is absorbed by the trees. The founder of Climate Care, Mike Mason, told the environment audit select committee in February: 'I think planting trees is mostly a waste of time and energy.' And yet Climate Care relies for some 20% of its online sales on forestry.

Mr Mason explained apologetically: 'People love it unfortunately.'

The idea of buying and retiring EU carbon permits is becoming equally discredited. The first phase permits, which run to the end of this year, are now worthless. The second phase, due to cap the carbon emissions of European companies from 2008 to 2012, are high-risk investments. Nobody knows whether the European commission has got its calculations right this time. Nor can anybody forecast the demand for carbon-heavy production, which will fluctuate with the weather and the economy. Andreas Arvanitakis, who monitors the market in these permits for the specialist analysts Point Carbon, would not use them for offsets: 'I have a completely green tariff, I offset my flights and I make sure that I am getting the absolute top stuff. I wouldn't touch this.'

Projects that use renewable energy or efficient energy to cut carbon are beset with the uncertainties of measurement and additionality. And many companies are selling speculative 'forward' credits: they have hooked up with some third-world project and started selling offsets on the assumption that the project will probably materialise and succeed.

Defra wants to rescue the market by introducing a voluntary standard. But its credibility is low. In January, it recommended four named companies, including Equiclimate with its worthless EU permits, and Carbon Offsets Ltd, which had not even started business at the time.

One measure of the crisis is the progress of the gold standard scheme, which is backed by Greenpeace and other environmental groups as a particularly rigorous process to ensure that emission reductions are verified, additional and consistent with sustainable development. When the scheme was launched three years ago, it was widely derided. Jasmine Hyman, the gold standard marketing director, said: 'The irony is that three years ago, we were defending our right to exist and everybody was saying, "Stop it with all your rules' and now we are the darling of the dance floor."

The scheme has registered seven projects, two of which have so far produced some 350,000 tonnes of verified gold standard carbon reductions. And suddenly, as so many other projects struggle with uncertainty, it has unfilled orders for eight million more.

16 June 2007
© *Guardian Newspapers Ltd, 2007*

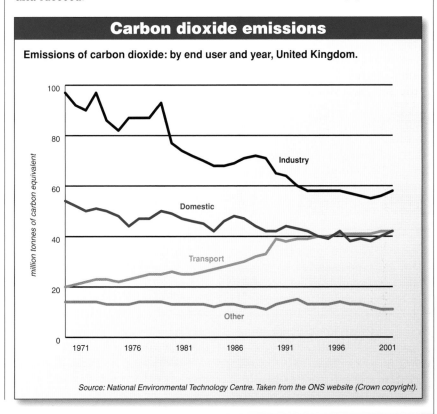

Carbon dioxide emissions

Emissions of carbon dioxide: by end user and year, United Kingdom.

million tonnes of carbon equivalent

Industry

Domestic

Transport

Other

Source: National Environmental Technology Centre. Taken from the ONS website (Crown copyright).

Muck and brass – with carbon credits

Animal waste – and the methane it produces – is just one way in which reducing emissions can be profitable, says Iain Dey

To an observer, it could seem as if the livestock farmers of Minas Gerais in south-eastern Brazil have suddenly become big fans of tennis. All over the countryside, large nylon domes that look like covered tennis courts have popped up in recent years. But the reality is even more curious.

These domes are part of the new global industry forming around the trading of carbon credits – the new global commodity that has emerged from the desire to cut the world's greenhouse emissions. Underneath the covers are large cesspits, filled with slurry created by pigs and cattle. The methane these pits generate is trapped, stopping it escaping into the atmosphere – and earning carbon credits which can then be sold on to European utilities, under pressure to meet EU emissions targets.

AgCert, a Dublin-based company listed on the London Stock Exchange, has funded more than 600 of these cess sites across Brazil and Mexico and is just one of a clutch of companies cashing in on this new commodity.

While these businesses have had only limited commercial success so far, a huge trading market has been created in carbon credits, fuelled by the European Union's emissions trading scheme which was launched in 2005. The first phase of that scheme descended into chaos after governments gave their industries too many credits, in effect allowing them to pollute more rather than less. EU ministers last week gave the impression that an upcoming EU summit on climate change would pave the way for the scheme to be extended further into the future, beyond the life of the Kyoto Protocol on climate change.

'Demand for carbon credits is growing faster than the rate at

which we can increase supply,' says Bill Haskell, the chief executive of AgCert, which has seen its share price halve since listing in June 2005. 'It's still a very young market, where the rules keep changing. But it looks like it's here to stay.'

Haskell is the first to admit that AgCert has made some mistakes in its time. That's not unusual among the companies trying to make money in the carbon-credit sector. Some companies have found their projects were far less effective than they had thought. One methane-trapping project in India met only about 10 per cent of its target carbon reductions. Others have been hit by delays.

Econergy International, which is generating credits from renewables projects across South America, has been hit by production delays which have knocked its shares down by about 14 per cent since listing a year ago.

Camco International, which is working on emission-reduction projects with a number of major steel producers in China and India, has seen its share price drop by about 30 per cent since listing last summer.

There are two main types of carbon credits in circulation – those awarded by the UN in line with the Kyoto agreement, and those distributed by the EU. The credits are awarded for projects, particularly in developing countries, that will help contribute to a sizeable shift in the mix of energy consumed in that area. The process is riddled with irony. For example, the methane pits in Brazil still emit sizeable amounts of greenhouse gases – but substantially less than would be created if the animal waste were allowed to decompose of its own accord. The credits are awarded for the net reduction effect.

Locating in the right area is also key – and can really make the economics work. Trading Emissions, another Aim-listed investment company specialising in carbon trading, is about to begin work on a $250m (£128m) hydro-electric power station in Peru. Its equity in the project will be about £23m. But because Peru's national grid relies heavily on coal, one of the dirtiest fuels around, the project gets lots of UN brownie points. It is expected to be credited with eliminating about 800,000 tonnes of carbon from the Peruvian air every year – which translates into carbon credits worth about £7m a year, at today's forward prices.

The EU carbon credits form the backbone of the carbon market and set the benchmark price for every carbon deal in the world. Since 2005 the EU has allocated an annual carbon allowance to every company running a sizeable power plant or generating greenhouse gases through steelmaking, oil refining or other industrial processes. Although utility companies and large industrial groups account for most of the market, even large schools and hospitals are included in the scheme.

In the first phase of the European scheme, which lasts until the end of this year, the allocations were set to encourage companies to cut their emissions by about 5 per cent. Those who exceeded the targets would be able to sell their surplus carbon allowance. Those who fell short would have to buy credits in the market – either EU credits or limited tranches of UN credits created by one of the carbon-generation companies.

The trading of carbon credits has proved to be the most lucrative end of the carbon business so far. According to Paul Newman, managing director of ICAP Energy, which handles about 40 per cent of the trades in the carbon market, between 500,000 and 1m tonnes of carbon are traded every day. That means that about 4 per cent of the credits issued by the EU are changing hands on a daily basis. Goldman Sachs and Morgan Stanley, the US investment banks, are among those to have put large sums of money in carbon-trading ventures. Proprietary trading desks at carbon producers such as BP, Shell and all the major European utility groups have also been trading heavily in the carbon market.

'There is a real market place for carbon credits, with no shortage of buyers and sellers, plenty of liquidity and real variations in price,' says Newman. 'Although this market technically only lasts until 2012 under the terms of the Kyoto agreement, I cannot imagine that it will not continue beyond that.'

One of the potential problems of extending the scheme is its structure. The first phase of the EU carbon system ends on 31 December. At that point all the carbon allowances issued under the first phase of the scheme become worthless.

Coal generates about twice as much carbon dioxide as gas

Carbon initially traded at about €7 (£4.70) a tonne, when the market opened in 2005. Last spring the price peaked at more than €30 a tonne, then crashed on evidence of oversupply in the market. Although forward prices for next year's credits were steady at around €13 a tonne last week, the current crop of credits were trading at less than €1 a tonne. At this price, there is no incentive to reduce emissions.

Coal generates about twice as much carbon dioxide as gas. One of the hidden rationales of the trading scheme is to encourage utilities to switch more power generation from coal to gas. With carbon credits as cheap as they are, it is more cost-effective to buy cheap coal, deliberately exceed the annual allocation, and make up the difference by buying credit in the market.

'It is fair to say that there is not a real incentive to stop using coal at the moment,' concedes James Wilde of the Carbon Trust, a government-funded company set up to help companies. 'But the key to this in future is to ensure that the right carbon allocations are awarded in the next phase of the scheme.'

The second phase of the EU carbon-trading scheme runs from 1 January next year until the end of 2012. Broadly speaking, the allocations will be cut – one UK utility executive said his company had been told to cut its emissions by 7 per cent in the first phase of the scheme, but expects it to look for reductions of between 10 per cent and 15 per cent. If the carbon-trading scheme is to work as an incentive, the price of carbon needs to increase. The final structure of the new market, when it emerges over the coming weeks, is likely to ensure that this is the case.

As the price increases and the market becomes more liquid, the number of parties making a cut from the process of turning Europe green will only increase.

24 February 2007
© *Telegraph Group Ltd, London 2007*

What are the causes of climate change?

Respondents were asked 'What do you think are the main causes of climate change?' (Top ten responses)

Cause	Youth (total)	Adults (March 06)
Emissions from cars/vans/buses	33%	26%
Air pollution	30%	31%[1]
Aerosols	26%	4%
Global warming	18%	5%
Destruction of rainforest, cutting down trees	16%	12%
Burning fossil fuels for energy	16%	12%
Using/wasting fuel in vehicles	13%	6%
Hole in the ozone layer	13%	4%
Carbon (dioxide) emissions	12%	15%
Greenhouse gases/greenhouse effect	12%	6%
Don't know	9%	13%

% 0   5   10   15   20   25   30   35

1. Adult respondents suggested 'pollution' – youth respondents were asked to specify the type (water, land, air).

Source: Attitudes to Climate Change – Youth Sample; June 2006. Prepared for DEFRA by COI. Crown copyright.

Positive energy

Harnessing people power to prevent climate change. Extract from a report by the Institute for Public Policy Research

Barely a week goes by without a press headline warning us of the dangers we face from climate change. Behind the stories, real people are already being hit, with climate change now killing 150,000 people a year. The technological solutions to prevent it from becoming much worse already exist. The challenge is to make the transition to them in time to avoid dangerous climate change.

Some of the changes needed to make that transition will be achieved entirely through regulations that largely affect industry. Others will require individuals to choose to behave differently. In the UK, the energy we use in our homes and for personal transport is responsible for 44 per cent of the country's carbon dioxide (CO_2) emissions. Engaging with the public is therefore critical to reducing the country's overall contribution to climate change.

Engaging the public will not only benefit the climate: helping individuals to use energy more efficiently and be less reliant on fossil fuels will also help government meet its other energy policy objectives of increasing energy security and reducing fuel poverty. More broadly, empowering people to exert control and resolve problems for themselves is a good in its own right: improving governance, deepening democracy and rebuilding trust.

When it comes to climate change, there is clear evidence that members of the public who are concerned about this issue do not always feel engaged in the societal challenge of tackling it, and feel locked into the systems and norms that fuel it. There is an urgent need to enable people such as these to act to reduce their contribution. The aim of this report is to find more effective ways of doing so.

Based on an extensive, cross-disciplinary literature review, interviews and a discourse analysis of UK climate change communications, this report suggests policies, techniques and communications approaches for promoting behaviour change. It is intended to help policymakers and others seeking to reduce the public's contribution to climate change to do so as effectively as possible.

Which behaviours need changing?

Almost 60 per cent of the contribution of an average UK citizen to CO_2 comes from using energy in the home. Of these emissions, three-quarters come from heating space and water alone (the single largest contributor

to emissions by individuals in a given year), and one-quarter from powering refrigerators, lights, ovens, washing and dishwashing machines, and consumer electronics.

Changes that will do the most to reduce individuals' CO_2 emissions from home energy use include fitting insulation in cavity walls and loft spaces, installing an efficient condensing boiler, and installing microrenewable technology for heat (such as solar thermal and biomass) and electricity.

Almost 60 per cent of the contribution of an average UK citizen to CO_2 comes from using energy in the home

---

The remaining 40 per cent of an average UK citizen's contribution to CO_2 comes from transport, including flying. Almost three-quarters of this can be attributed to car use, with almost a quarter coming from flying.

Consequently, changes that will do most to reduce individuals' transport emissions include cycling, walking, using public transport, buying lower-carbon cars (such as those with smaller or hybrid engines, or that use biofuels), and driving more efficiently ('eco-driving'). Flying less, by taking holidays nearer to home or by train, or offsetting flights effectively, will also help.

People can also affect climate change through less direct means, such as purchasing food and consumer products that have been made using less energy and transported smaller distances, or taking part in campaigning to encourage decision-makers to take action on climate change. Each of these is a legitimate, and potentially valuable, avenue of individual action.

However, in most cases it is still very difficult to assess accurately the significance of such actions. As a result, this report focuses mainly on those changes in behaviour relating to energy use in the home and in transport, whose contribution to climate change through emissions can be easily measured.

What is the public doing about climate change?

It is clear from the evidence that the majority of people in the UK are not taking many actions to mitigate their emissions in a significant way.

A large number of homes are still not properly insulated. Almost two-thirds (63 per cent) of homes that could have cavity-wall insulation – some 8.3 million homes – have not installed it. Similarly, in 2003, 48 per cent of homes that could have had loft insulation fitted at the optimum depth (4 inches) did not do so.

Since 2005 it has been mandatory to fit the most efficient type of condensing boiler. However, there are a further 15 million that need to be replaced. Homes are also kept appreciably warmer than they were 30 years ago: between 1990 and 2004 there was a rise of 1.9°C in internal temperatures.

While people do increasingly buy energy-efficiency A-rated appliances, the energy savings from doing so have been more than offset by the 50 per cent growth in the number of appliances in the home between 1990 and 2004 – especially in consumer electronics. Exacerbating the problem further, some new products consume more electricity than the products they replace (for example, plasma televisions consume 4.5 times more energy than their cathode ray tube predecessors).

By contrast, investment in micro-renewables is still a tiny niche market, as is 'green' tariff electricity (electricity that energy companies have produced from renewable sources of energy). There are currently only around 100,000 microgeneration installations in the UK, representing under 0.4 per cent of UK households. And just 212,000 customers have switched to a green electricity tariff, representing some 0.83 per cent of the total electricity market in 2005.

> **Although there has been an increase in public transport use, it still makes up only 8 per cent of the total number of trips made**

A similar picture exists for transport choices. People are using their cars to travel further, and more often, with an 18.5 per cent increase in the number of vehicle kilometres by cars and taxis since 1990. Car ownership is also increasing: there were nearly one-third more cars on UK roads in 2005 than in 1990 – equivalent to another 7.5 million more cars.

Although there has been an increase in public transport use, it still makes up only 8 per cent of the total number of trips made. Only London has seen a shift away from car use to buses and an increase in cycling.

Outside of the capital, local bus use has declined on average by almost 12 per cent since 1990. Nationally, cycling represents only 1.5 per cent of all journeys made and the distance travelled and number of trips taken by bicycle have fallen by 6 and 20 per cent respectively. Participation in car clubs and car-share schemes remains a niche choice.

The one area of positive change is that average emissions from the nation's car fleet are coming down. However, this is largely being driven by technology rather than by consumer behaviour. In 2005, total sales of low-carbon vehicles (LCVs), which are mostly hybrid cars, amounted to just 0.3 per cent of the market.

Meanwhile, few motorists recognise the concept of driving more efficiently. Data from average vehicle speeds on motorways shows that a majority (56 per cent) of cars exceed the 70 miles per hour speed limit, with more than one-third of drivers exceeding 80 miles per hour.

Lastly, air travel is now more popular than ever, and offsetting remains a small minority practice. Between 1994 and 2004 the number of passengers flying abroad from the UK rose by about 65 per cent, and the number flying domestically by about 70 per cent. According to a 2006 poll, just 1-5 per cent of respondents said they offset their emissions from flying.

⇨ The above information is reprinted with kind permission from the Institute for Public Policy Research. Please visit www.ippr.org for more information.

© IPPR 2007

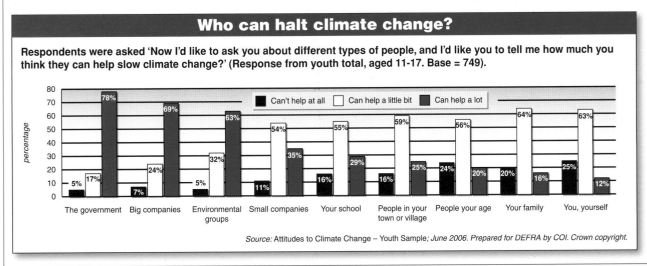

Who can halt climate change?

Respondents were asked 'Now I'd like to ask you about different types of people, and I'd like you to tell me how much you think they can help slow climate change?' (Response from youth total, aged 11-17. Base = 749).

Legend: ■ Can't help at all □ Can help a little bit ■ Can help a lot

	Can't help at all	Can help a little bit	Can help a lot
The government	5%	17%	78%
Big companies	7%	24%	69%
Environmental groups	5%	32%	63%
Small companies	11%	54%	35%
Your school	16%	55%	29%
People in your town or village	16%	59%	25%
People your age	24%	56%	20%
Your family	16%	64%	20%
You, yourself	12%	63%	25%

y-axis: percentage

Source: Attitudes to Climate Change – Youth Sample; June 2006. Prepared for DEFRA by COI. Crown copyright.

KEY FACTS

⇨ People are causing the change by burning nature's vast stores of coal, oil and natural gas. This releases billions of tonnes of carbon dioxide (CO_2) every year, although the changes may actually have started with the dawn of agriculture, say some scientists. (page 1)

⇨ The world is becoming more humid under climate change and exacerbating global warming, research reveals. (page 2)

⇨ It is true that the world has experienced warmer or colder periods in the past without any interference from humans. The ice ages are well-known examples of global changes to the climate. However, the increase of three-quarters of a degree centigrade (0.74°C) in average global temperatures that we have seen over the last century is larger than can be accounted for by natural factors alone. (page 3)

⇨ Change in solar activity is one of the many factors that influence the climate but cannot, on its own, account for all the changes in global average temperature we have seen in the 20th century. (page 5)

⇨ There are real concerns that, in the long term, rising levels of greenhouse gases in the atmosphere could set in motion large-scale and potentially abrupt changes in our planet's natural systems and some of these could be irreversible. (page 6)

⇨ 'Greenhouse gases' are so called because they act like the glass of a greenhouse, trapping heat from the sun to warm up the Earth. Most of these gases occur naturally and with-out them our planet would be too cold to sustain life, but the balance is a very delicate one. (page 7)

⇨ The world is warming faster than at any time in the last 10,000 years. (page 8)

⇨ Climate change will increasingly drive biodiversity loss, affecting both individual species and their ecosystems. An ecosystem can be defined as a community of plant and animal species and the physical environment that they occupy, which includes the climate regime. (page 9)

⇨ 18% of respondents in a 2007 Ipsos MORI poll agreed that human activity does not have a significant effect on the climate. 69% did not agree. (page 12)

⇨ Collectively the G8 – Britain, France, Germany, Italy, Canada, USA, Japan and Russia – which represent just 13% of the world's population, are responsible for around 43% of the world's greenhouse gas emissions. (page 16)

⇨ 40% of respondents in a 2007 Ipsos MORI poll felt that recycling would do the most to help reduce climate change, followed by developing cleaner engines for cars (34%) and avoiding creating waste in the first place (22%). (page 19)

⇨ Our everyday actions consume energy and produce carbon dioxide emissions, for example driving a car, heating a home or flying. Offsetting is a way of compensating for the emissions produced with an equivalent carbon dioxide saving. In this way it lessens the impact of a consumer's actions. (page 20)

⇨ Forest ecosystems make an important contribution to the global carbon budget. This is because of their potential to sequester carbon in wood and soil but also because of their potential to release it if forests are cleared. (page 26)

⇨ The solutions to climate change exist, primarily in curbing the consumption of fossil fuels in the developed world, and in technologies which harness unlimited and unpolluting renewable energy. (page 27)

⇨ The amount of carbon dioxide (CO2) in the atmosphere has been rapidly increasing over the last 100 years. As the proportion of CO2 in the atmosphere changes, the way it retains heat also changes. Scientists now believe this is what is causing climate change. (page 30)

⇨ A radical form of 'offsetting' carbon dioxide emissions to prevent climate change has been proposed – having fewer children. Each new UK citizen less means a lifetime carbon dioxide saving of nearly 750 tonnes, a climate impact equivalent to 620 return flights between London and New York, the Optimum Population Trust says. (page 31)

⇨ 53% of respondents in a 2007 Ipsos MORI poll agreed that they would do more to try and stop climate change if other people did more, too. 33% did not agree with this statement. (page 31)

⇨ New small zero carbon 'eco-towns' built on brownfield land could lead the way in cutting carbon emissions and building affordable homes. (page 32)

⇨ 90% of young people aged 11 to 17 surveyed by DEFRA were aware of climate change, compared to 95% of adults surveyed. (page 34)

⇨ Climate change now kills 150,000 people a year. (page 38)

⇨ In the UK, the energy we use in our homes and for personal transport is responsible for 44% of the country's carbon dioxide (CO_2) emissions. (page 38)

GLOSSARY

Biosphere
The part of the earth and atmosphere in which life exists.

Conjecture
A theory or opinion based on guesswork.

Discourse
Communication exchanges, usually writing or talking about a particular topic.

Discrepancy
An unexpected or unexplained difference in results.

Exacerbate
To make worse.

Facilitate
To make easier or to help.

Fertiliser
A nutrient added to soil to help plants grow.

Hydrology
The scientific study of water.

Inexhaustible
Unable to be used up; a never-ending supply.

Liquidity (market)
Being able to convert assets into cash without significantly affecting the assets' value.

Magnitude
The size or extent of something.

Mandatory
Compulsory or required by law.

Mitigation
Action taken to reduce the harm or severity of something.

Peripheral
Near the outside or edge.

Permafrost
A layer of permanently frozen soil underground.

Positive feedback
Feedback that amplifies or reinforces a process.

Retrofitted
Changes to the design or structure of an existing object.

Scepticism
Doubt or disbelief.

Sediment
Material that has been transported and deposited by water or wind flow.

Succour
Relief or assistance in times of difficulty.

Terrestrial
Living on land.

INDEX

Additional Resources

Other Issues *titles*

If you are interested in researching further some of the issues raised in *Climate Change*, you may like to read the following titles in the **Issues** series:

⇨ Vol. 150 *Migration and Population* (ISBN 978 1 86168 423 3)

⇨ Vol. 146 *Sustainability and Environment* (ISBN 978 1 86168 419 6)

⇨ Vol. 140 *Vegetarian and Vegan Diets* (ISBN 978 1 86168 406 6)

⇨ Vol. 138 *A Genetically Modified Future?* (ISBN 978 1 86168 390 8)

⇨ Vol. 134 *Customers and Consumerism* (ISBN 978 1 86168 386 1)

⇨ Vol. 119 *Transport Trends* (ISBN 978 1 86168 352 6)

⇨ Vol. 111 *The Waste Problem* (ISBN 978 1 86168 344 1)

⇨ Vol. 110 *Poverty* (ISBN 978 1 86168 343 4)

⇨ Vol. 109 *Responsible Tourism* (ISBN 978 1 86168 329 8)

⇨ Vol. 97 *Energy Matters* (ISBN 978 1 86168 305 2)

⇨ Vol. 76 *The Water Crisis* (ISBN 978 1 86168 265 9)

For more information about these titles, visit our website at www.independence.co.uk/publicationslist

Useful organisations

You may find the websites of the following organisations useful for further research:

⇨ **Department for Environment, Food and Rural Affairs (DEFRA)**: www.defra.gov.uk

⇨ **edie**: www.edie.net

⇨ **International Institute for Environment and Development (IIED)**: www.iied.org

⇨ **Met Office**: www.metoffice.gov.uk

⇨ **New Scientist:** http://environment.newscientist.com

⇨ **OneWorld**: www.oneworld.net

⇨ **People & Planet**: http://peopleandplanet.org

⇨ **Royal Society**: www.royalsoc.ac.uk

⇨ **Science Museum**: www.sciencemuseum.org.uk

⇨ **Stop Climate Chaos**: http://icount.org.uk

⇨ **United Nations Environment Programme World Conservation Monitoring Centre**: www2.wcmc.org.uk

ACKNOWLEDGEMENTS

The publisher is grateful for permission to reproduce the following material.

While every care has been taken to trace and acknowledge copyright, the publisher tenders its apology for any accidental infringement or where copyright has proved untraceable. The publisher would be pleased to come to a suitable arrangement in any such case with the rightful owner.

Chapter One: Our Changing Climate

Instant expert: climate change, © New Scientist, World becoming more humid, © Crown copyright is reproduced with the permission of Her Majesty's Stationery Office, Climate change controversies, © Royal Society, Why it's green to go vegetarian, © Vegetarian Society, What does climate change mean for us?, Stop Climate Chaos, Biodiversity and climate change: ecosystems, © United Nations Environment Programme World Conservation Monitoring Centre.

Chapter Two: Climate Politics

Climate change and cities, © International Institute for Environment and Development, Climate change denial, © George Monbiot, The deceit behind global warming, © Telegraph Group Ltd, G8 climate change accord elicits mixed reactions, © OneWorld.net, UK legislation: Climate Change Bill, © Crown copyright is reproduced with the permission of Her Majesty's Stationery Office, Climate change is like 'World War Three', © Telegraph Group Ltd, Right to be suspicious, © Guardian Newspapers Ltd, Carbon offsetting – frequently asked questions, © Crown copyright is reproduced with the permission of Her Majesty's Stationery Office, Rockin' all over the world, © Guardian Newspapers Ltd.

Chapter Three: Climate Solutions?

Can algae save the world?, © Board of Trustees of the Science Museum, Oceans offer climate cure, © edie, Carbon sequestration, © Crown copyright is reproduced with the permission of Her Majesty's Stationery Office, Fuelling the future, © People & Planet, Sustainable fossil fuels, © Simon Fraser University, Climate change and the need for renewable energy, © BWEA, Combat climate change with fewer babies, © Optimum Population Trust, New eco-towns could help tackle climate change, © Crown copyright is reproduced with the permission of Her Majesty's Stationery Office, The inconvenient truth about carbon offsetting, © Guardian Newspapers Ltd, Muck and brass – with carbon credits, © Telegraph Group Ltd, Positive energy, © IPPR.

Photographs

Flickr: pages 13 (John LeGear); 17 (Daniel Tomczak); 22 (Karl Sinfield); 33 (Andrew*); 38 (mandamonius).
Stock Xchng: pages 3 (B S K); 26 (Martin Boulanger); 29 (Craig Jewell).

Illustrations

Pages 1, 15: Simon Kneebone; pages 8, 23: Don Hatcher; pages 11, 36: Bev Aisbett; pages 18, 28: Angelo Madrid.

Additional research and editorial by Claire Owen on behalf of Independence Educational Publishers.

And with thanks to the team: Mary Chapman, Sandra Dennis, Claire Owen and Jan Sunderland.

Cobi Smith and Lisa Firth
Cambridge
January, 2008

Maths
Problem
Solving

**AGES
7-11**

JOHN DABELL

Author
John Dabell

Editor
Sally Gray

Assistant Editor
Linda Mellor

Series Designers
Anthony Long and
Joy Monkhouse

Designer
Allison Parry

Illustrations
Matt Ward, Beehive Illustration

Photographs
Derek Cooknell

The publishers would like to thank:

Gerry Bailey and the staff and pupils at
**Clapham Terrace School,
Leamington Spa**.

Published by Scholastic Ltd
Book End
Range Road
Witney
Oxfordshire OX29 0YD

www.scholastic.co.uk

Text © 2006 John Dabell
© 2006 Scholastic Ltd

Designed using Adobe InDesign

Printed by Bell and Bain Ltd., Glasgow

7 8 9   0 1 2 3 5

British Library Cataloguing-in-Publication Data

A catalogue record for this book is available from the
British Library.

ISBN 0-439-96570-5
ISBN 978-0439-96570-5

Mixed Sources
Product group from well-managed
forests and other controlled sources
www.fsc.org Cert no. TT-COC-002769
© 1996 Forest Stewardship Council
FSC